HOW PLANTS WORK

HOW PLANTS WORK

The science behind the
amazing things plants do

LINDA CHALKER-SCOTT

TIMBER PRESS

PORTLAND · LONDON

Frontispiece: My home landscape.
Page 6: My garden pond.

Photo and illustration credits appear on page 222.

Thanks are offered to those who granted permission for use of photos. While every reasonable effort has been made to contact copyright holders and secure permission for all materials reproduced in this work, we offer apologies for any instances in which this was not possible and for any inadvertent omissions.

The information in this book is true and complete to the best of our knowledge. All recommendations are made without guarantee on the part of the author or Timber Press. Mention of trademark, proprietary product, or vendor does not constitute a guarantee or warranty of the product by the publisher or author and does not imply its approval to the exclusion of other products or vendors.

The Haseltine Building
133 S.W. Second Avenue, Suite 450
Portland, Oregon 97204-3527
timberpress.com

6a Lonsdale Road
London NW6 6RD
timberpress.co.uk

Printed in China

Text design by Stacy Wakefield Forte
Cover design by Anna Eshelman

Library of Congress Cataloging-in-Publication Data

Chalker-Scott, Linda.
How plants work: the science behind the amazing things plants do/Linda Chalker-Scott.—First edition.
        pages cm
Includes index.
ISBN 978-1-60469-338-6
1. Plant physiology. 2. Gardening. I. Title.
QK711.2.C425 2015
571.2—dc23   2014032414

A catalog record for this book is also available from the British Library.

In December 2013, as I was finishing this book, I lost my dad to cancer. He was a structural engineer, and while my intellectual interests led in other directions, I got my passion for discovering how things work from him. This book is dedicated, with much love, to Raymond Lloyd Chalker.

CONTENTS

From the Ground Up

MY FIRST HOUSE WAS IN CORVALLIS, Oregon, home of Oregon State University, where my husband and I were working on our Ph.D. research in horticulture. Our tiny front yard had enough room for a single specimen tree, in this case a lovely 'Bloodgood' Japanese maple *(Acer palmatum* var. *atropurpureum)*. We were eager to update the front entry, and we replaced the 1930s era concrete steps with a basket-weave brick entry and wooden deck. The design perfectly showcased our maple, and we were thrilled with the transformation.

Well, until the next year. Suddenly, our maple tree didn't grow so well. Many of the branches died. Finally, it was in such dire straits that we dug it up and replaced it with a much smaller tree, which thrived. But what happened to the first tree?

I asked one of my favorite professors at Oregon State about the sudden demise. Jim Green was our department's extension specialist and didn't teach any of my graduate classes. But he was knowledgeable, easy-going, and had a wicked sense of humor. The graduate students loved him.

Imagine my shock, then, when he turned visibly angry as I explained our landscaping changes and subsequent tree death.

"What in the world did you think would happen," he snapped, "when you disturbed seventy-five percent of the tree's root zone in the middle of summer?"

Wow. I hadn't even thought about that. We'd dug up the existing lawn and laid down bricks and deck timbers. I remember silently cursing the roots as we dug. And they were probably cursing us back. I felt stupid, not just because I had irked Jim, but because I hadn't foreseen these consequences myself. After all, I was getting a Ph.D. in horticultural plant physiology!

In hindsight, I think this was a defining moment for me, though I didn't know it at the time. I do know that it was during this time I became more curious as to how plants responded to different environmental stresses (besides dying, of course). Over three decades I evolved from a laboratory plant physiologist (studying how plants function and interact with their environment), to an applied urban horticulturist, and finally to an extension specialist at Washington State University. Though my career continued to change, my interest in how plants work only became more engrossing. I combed through articles on soil science, arboriculture, environmental horticulture, and restoration ecology as well as those in the more traditional botany and horticulture journals.

While botany books describe leaves and roots in isolation from each other (and therefore have chapters called "The Leaf" or "The Root"), physiology is the study of systems. It's impossible to explain the physiology of a leaf or a root, because their functions are influenced by other plant parts. It would be like describing how a heart works without mentioning the lungs that provide the oxygen or the arteries and veins that deliver and return blood. Instead, plant physiology textbooks have chapters on photosynthesis, mineral uptake, and flowering. However, it can be difficult to make these conventional topics both accessible and interesting to the nonscientist. Current books on plant physiology are primarily focused on newer research at the molecular and genetic levels. The content is timely and important, but boy does it turn off your average gardener, who probably sees no practical connection between gene regulation in corn and the corn growing in the backyard vegetable garden. What we gardeners most want to know is how plants work so that we can have gardens and landscapes that are healthy, beautiful, and don't need constant additions of fertilizers and pesticides.

So this book is structured a bit differently. Each chapter opens with a real-life situation, often something in my own garden that I invite you to explore with me. Then I integrate the science as needed to answer questions that gardeners invariably have. I've tried to include all of the practical topics that you would find in a textbook, with examples and illustrations to make the science useful and easily understandable.

Understanding how plants work helps
gardeners design and manage landscapes that
require less fertilizer and fewer pesticides.

What kinds of things will you discover? We begin at the microscopic level and explore the basic machinery of plant cells, which is
substantially different from that of our own cells. Next we explore a
vital plant network hidden underground, the root system, which in
conjunction with fungal partners seeks out water and nutrients. In
chapter 3, we consider which minerals are essential for plant survival,
growth, and reproduction and which can be detrimental to these processes. Plants have the unique ability to combine water and carbon

dioxide in leaves to produce their own food, and photosynthesis is the topic of chapter 4.

Many gardeners live in seasonal climates, and our deciduous trees put on a wonderful display of reds, oranges, and yellows in autumn. In chapter 5, we'll discover that the plant pigments behind these colorful displays also can help plants retain water, live in salty or contaminated soils, avoid freeze damage, and fight diseases. Yet another pigment is the focus of chapter 6, which describes how plants measure day length to determine when it's time to germinate, flower, drop their leaves, and close up shop for the winter.

We think of plants as sedentary creatures, but in fact they move quite a bit. In chapter 7, we'll learn how plants move to follow the sun, avoid predators, and even capture food. Unfortunately, these movements sometimes put plants into places where we don't want them, so we grab the pruning shears. Chapter 8 explores how plants respond to pruning, staking, and other forms of manipulation that we try to impose on them. The final chapter investigates every plant's ultimate goal: leaving behind offspring to carry their genes forward. Plants produce an amazing array of pigments, fragrances, and seed structures that help them manipulate their environment and pollinators—including gardeners—to achieve this goal.

Throughout the book, I've included advisory sidebars on which gardening products and practices work and which don't. You'll find that many of the products and practices you've sworn by for years are not only a waste of money, but may actually harm your plants and soil. Just so you know, I used to buy the same products and follow the same practices, even with my training in horticulture. For instance, did you know that phosphate fertilizer can make plants work unnecessarily to take up water and nutrients? Or that the biggest barrier to getting your new tree to establish is amending the backfill soil with organic material? Think that Epsom salts will nourish plant roots just like they do your feet? Better think again! Understanding how plants work will help you predict what garden products are worth a try and which are best left on the shelf.

We often make gardening more of a chore than it needs to be, making decisions about plant care based on how we think plants will respond. Everything from watering to fertilizing to pruning or mowing is dictated by this mindset. Unfortunately, many gardeners make decisions that aren't just erroneous but may cause actual and long-lasting harm to plants, soils, and the surrounding environment. When you know how plants work, you'll understand how to use natural processes to your benefit. For instance, proper mulching drastically reduces weeds. Pruning at the right time and in the right place reduces explosions of unwanted growth that have to be pruned again. By using natural plant responses to nurture your garden or to outfox weeds, you'll have more time to spend watching and learning from your garden, rather than constantly fighting it.

I hope you'll find yourself rereading this book as you explore your own garden with newfound curiosity and fascination.

Let's get started!

Under the Microscope

IT'S TAKEN NEARLY FIFTEEN YEARS, but our garden is finally something I'm happy to share with visitors. In the front yard, a lovely pond dominates the sunny landscape, surrounded by small trees, shrubs, and groundcovers in every imaginable foliar shade: black mondo grass, lime green ginkgo, brick red laceleaf maple, and a contorted larch that turns the most amazing fiery colors in autumn. Following the flagstone pathway around the house and through a narrow glade filled with rhododendrons and azaleas, we enter the shady, north-facing backyard with its fragrantly flowered sarcococca and hardy gardenias. Ferns and mosses, ancient plants, are at home in this environment, as are the hostas with their dinner-plate-sized leaves. Hydrangeas thrive, untouched by pests or disease. On the deck in a sunny spot is my bog garden, a large ceramic container full of carnivorous pitcher plants and sundews.

The curious gardener might wonder why there are so many different leaf colors. What makes those flowers smell so good? How can a tree grow in bizarre contortions and maintain that shape? And why do some plants successfully avoid pests and disease? To answer these and many other questions that gardeners have asked for centuries, we need to step out of the garden and into the laboratory to explore the world of plant cell biology.

This chapter gives you a quick and selective introduction to plant anatomy and biochemistry. There are some structures, chemicals, and processes associated with plant cells that we need to discuss before we can look at how a plant works in its environment. First, we'll look at several cell structures: some that are found in both plants and animals and others that are unique to plants. Next, we'll consider the vast pharmacy of biochemicals manufactured by plants to fabricate, communicate, and regulate. Finally, we'll look at three specialized cells and tissues whose activities we can help—or hurt—when we're gardening.

Plant cells are connected to one another by plasmodesmata.

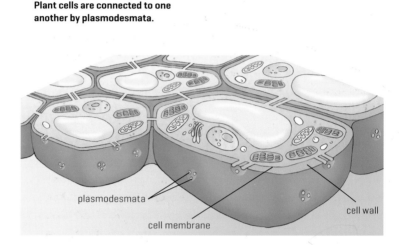

plasmodesmata

cell wall

cell membrane

The Cell Wall: Home Protection

Unlike animal cells, plant cells are surrounded by a rigid wall made of cellulose and other biochemicals. Rigid cell walls ultimately allow trees to grow to great heights and protect individual cells from environmental injury and biological attack. But while the cell wall protects the cell and gives it rigidity, it also isolates cells from one another, which makes transport of materials between cells very difficult. Fortunately, the walls are pierced by tiny holes through which the inner cell membranes extend. These tiny membrane tubes, called plasmodesmata, are an important means of transport and communication between cells. Plasmodesma (the singular form) has a very cool Greek meaning: *plasm* refers to form and *desma* is a girl's name meaning binding oath. So plasmodesmata bind forms together, in this case plant cells.

Individual cells are also cemented to one another by a gelatinous layer of pectins, carbohydrates that you're familiar with if you make preserves. Without pectin, jams and jellies would be more like syrup, as would young plant tissues.

Why do plants have cell walls, rather than some kind of internal skeleton? Plant cells have to withstand internal water pressure, and without the wall, the cell membrane would burst. This water pressure is vital in moving sugars, minerals, gases, and other dissolved substances throughout the plant.

The Cell Membrane: Border Patrol

Both plant and animal cells are surrounded by a membrane, which keeps all the contents floating together in a soupy liquid called cytoplasm. The cells have various organelles (literally, little organs) that have specific tasks, such as producing energy (mitochondria), directing cell reproduction (nucleus), or packaging materials (Golgi apparatus and Golgi bodies). Nutrients, dissolved gases such as oxygen and carbon dioxide, and building blocks for various compounds are all contained in the cytoplasm. Besides keeping all the organelles and cytoplasm contained, the cell membrane plays a critical role in determining what's allowed to enter and leave the cell.

As an analogy, let's consider the border between any two countries. I travel frequently between Washington State and British Columbia, and when I enter either country a customs official questions me about fruit, alcohol, and other materials I might be transporting. Likewise, receptors on the cell membrane inspect various molecules to determine whether they should be allowed into or out of the cell. Usually this system works well, but occasionally something unwanted might slip across, like a virus or an environmental pollutant.

The Mitochondria: Power Plants

Those of you who remember your basic high school biology might recall this nifty alliteration: "the mighty mitochondria are the powerhouses of the cell." All plant and animal cells contain mitochondria.

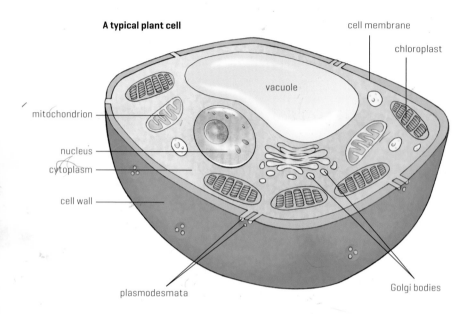

A typical plant cell

cell membrane

chloroplast

vacuole

mitochondrion

nucleus

cytoplasm

cell wall

plasmodesmata

Golgi bodies

Here's the interesting thing: the mitochondria are, in fact, alien invaders. Mitochondria contain DNA that's radically different than the DNA in the cell's nucleus, and they reproduce independently of the cell's reproductive cycle. (Fun fact: the mitochondria in most plants and animals are derived from those in the egg. That's how maternal lines are established.) Mitochondrial DNA is most similar to that found in present-day bacteria.

How could this possibly have happened? The most convincing theory, which is supported by a great deal of evidence, is that ancient, single-celled ancestors of plants and animals engulfed ancient bacterial cells. This isn't as far-fetched as you might think. Going back to high school biology again, you might remember watching a paramecium or amoeba (single-celled animals) gobble up little algal or bacterial cells under a microscope. The theory states that some of these bacterial cells were not digested and instead became permanent cellular guests. Because they benefitted the host cell by providing energy, the descendants of the original host cells retained the bacteria, which eventually lost the ability to live independently and are now cell organelles.

The Chloroplasts: More Alien Invaders

One of the obvious differences between plants and animals is that plants photosynthesize using chlorophyll-rich chloroplasts to capture light energy and convert it to sugar. Like mitochondria, chloroplasts contain unique DNA and replicate independently. However, chloroplast DNA is most similar to that found in blue-green algae. So plant cells contain two different alien invaders, both of which benefit the host cell by producing energy.

When plant cells eventually divide, the nucleus orchestrates the action while mitochondria and chloroplasts simply go along for the ride. It's not an equal partitioning, so dividing cells usually have different numbers of chloroplasts in each half. From a gardener's perspective, this is where some really interesting things can happen in plants.

Just like nuclear DNA, chloroplast DNA also collects mutations (spontaneous, random changes in the genetic code). Some of these mutations affect the color of the chloroplast itself. If a newly dividing cell only contains white chloroplasts, for instance, then that cell will be white, as will all of its descendants. What started as a single white cell becomes a white blotch and eventually a variegated leaf.

The Vacuole: Warehouse Central

When we look at a plant cell under the microscope, most of it appears to be made up of nothing. That nothing is the vacuole (meaning vacuum). Because plants don't have an excretory system like animals do, for many years vacuoles were thought to be cellular garbage dumps. The vacuole is a membrane filled with water and lots of dissolved substances: acids, sugars, and an amazing assortment of what are collectively called secondary compounds. In a healthy cell, the vacuole is so bloated that it may take up 95 percent of the entire cell, pressing all the other organelles against the cell membrane. Oddly, this is actually good for cell health, because oxygen, light, and other necessities enter the cell through the membrane.

I don't like crowds, and I really don't like crowds in small spaces, like elevators. When I'm in a hot and stuffy crowded elevator, I try to stand along a wall, especially near the front where fresh air rushes in when the doors open. So, although it might seem problematic that the chloroplasts, mitochondria, and other organelles are squashed against the cell membrane by the vacuole, it's the best possible place for them to be to get oxygen, light, and other resources.

The vacuole is able to absorb and retain all of this water because it stores a lot of dissolved substances. An important mantra in understanding how plants work is "Water always moves to where water isn't." The more compounds the vacuole accumulates, the less space is left for water. This imbalance attracts more water into the vacuole, which creates high turgor pressure inside the cell. When the vacuole simply can't expand any more—when it squashes into the cell wall—then water flow stops.

What about all of those dissolved compounds that vacuoles supposedly store? Are they really waste products, as once thought? Or do they have more important roles to play?

Secondary Compounds: A Plant's Personal Pharmacy

Many of the dissolved substances within the vacuole are secondary compounds. This name is unfortunate because it suggests these chemicals are not as important as primary compounds. While primary compounds are those required for plant growth and development (things like sugars, DNA, fats, and proteins), various secondary compounds have defensive, attractive, or as yet unknown properties. So, even if they aren't required for basic growth, secondary compounds are crucial in a plant's ability to compete and survive in its environment.

I can still remember the "wow" moment when I learned in my graduate plant biochemistry class that we probably have yet to discover 90 percent of the compounds that plants make. I wouldn't be surprised if the number was closer to 99 percent. Researchers have done a good job of investigating secondary compounds that have economic benefit, from vanilla to pyrethrin to codeine, but it's likely that every plant species makes one or more unique compounds. We haven't even scratched the surface.

Secondary compounds are usually classified based on the biochemical pathway through which they are constructed by the plant (such as the alkaloids, phenolics, and terpenoids). This does little to help gardeners (or anyone else) understand what roles they play in a plant's life. Rather than turn this into a side trip through Biochemistry Land, let's look at the major roles secondary compounds play in plant survival.

BUILDING BLOCKS

LIGNIN. Every time you get a splinter, struggle to cut through a tree branch, or marvel at the ability of a spirally trained spruce to maintain its shape, you can thank (or blame) lignin. This complex compound toughens up the walls of mature cells and allows them to resist many environmental stresses, pests, and disease. It is slow to decompose, which gives woody mulch its staying power.

CUTIN. Cutin is part of the cuticle, which is the flexible, waxy layer protecting leaves and other soft tissues above ground. Cutin and other associated biochemicals help keep water in and invaders out of leaves. Its waxy nature is sensitive to soap, however, which is why many herbicides contain a soapy substance that helps breach the cuticle and deliver death to weeds.

SUBERIN. While leaves are covered in cutin, roots are swathed in a suberin sheath that has the same protective function. Additionally, suberized cells help select what substances in the roots will be transported through the xylem into the rest of the plant. The suberin in bark is what makes bark mulch a pretty poor choice, as the waxiness repels water absorption and movement.

CAROTENOIDS. Like their name suggests, carotenoids are carroty-orange pigments, although they can range in color from pale yellow to nearly red. Though their presence in leaves is masked by chlorophyll, carotenoids are important in scavenging the green light that chlorophyll can't capture for the photosynthetic machine.

ENVIRONMENTAL AND DISEASE PROTECTION

ANTIOXIDANTS. Many secondary compounds have antioxidant activity, including carotenoids and anthocyanins, the pigments that make plant tissues orange or various shades of red, blue, and purple, respectively. In plant cells, antioxidants absorb and neutralize reactive chemicals that otherwise could damage sensitive membranes. We get at least some of these benefits when we eat pigment-rich plant parts, and it certainly makes me feel virtuous about my fresh raspberry addiction (especially when they're out of season locally).

SUNSCREEN. Some compounds, like anthocyanins, absorb excess solar energy so leaves aren't overloaded. Others, like cuticular waxes, reflect sunlight and reduce the heat load. Plants with gray-green or blue-gray foliage are loaded with these waxes. Indoor and greenhouse-grown plants don't waste energy on these protective agents, however, and they will fry just like my Irish skin does if they're moved outside into full sun. Tuck them into shaded areas for several days first to let them toughen up.

ANTIMICROBIAL AGENTS. When attacked by fungal pathogens, plants can use existing saponins, soap-like compounds that break down fungal cell walls. Some other chemicals, called phytoalexins, are made only when pathogens attack. Finally, lignin and other structural

LEFT: **Plants from hot environments, like this manzanita, often have gray-green foliage to reflect excess sunlight.** RIGHT: **The blue-gray foliage of this palm is adapted to sunny conditions.**

compounds can be used as cellular reinforcements to wall off the rest of the plant from disease.

HERBIVORE PROTECTION

Many of the secondary compounds produced by plants to fend off herbivores will repel, deter, or kill you as well. When plants are domesticated for human consumption, we breed many of these defensive compounds out of them. That's why your fruits and veggies are so inviting to insects and other pests: they no longer have their natural abilities to keep predators away.

Harpin

THE PRODUCT

Sold as Messenger, harpin is derived from a protein isolated from fire blight bacteria (*Erwinia amylovora*).

THE SUPPOSED BENEFITS

Harpin acts like a vaccination to prepare plants to fight environmental stress and disease.

HOW PLANTS RESPOND

When it enters a plant cell, harpin turns on biochemical pathways that produce anti-stress compounds. It's a process that works great in the laboratory, but has little success with real plants in the landscape. The reason is pretty easy to understand. Remember that cell wall and those layers of waterproofing compounds on the leaf surface? It's really difficult to force the harpin molecule through the cuticle and cell wall. If harpin can't get into the cell, it can't turn on the necessary biochemical pathways. This is a product that's a good idea, but with limited practical use.

A protein from the bacterium responsible for fire blight helps plants fight disease in the laboratory, but not in the garden.

REPELLANTS. The odor of these compounds drives plant-eating insects and animals away before they can even take a bite. We take advantage of aromatic repellants, like essential oils of clove, cinnamon, and lemon, as natural pesticides. How many times do you see insect damage on mint leaves as you're picking them to make jelly or juleps? Obviously, like us, other animals are also attracted to rather than repelled by these compounds.

DETERRENTS. The taste of these compounds keeps snacking herbivores from coming back for seconds. These are often acidic, bitter, or astringent: think Mr. Yuk. Tannins, named for their early use in tanning leather, are good examples of astringent deterrents. When you take a mouthful of a nice dry red wine or bite into a too-green banana, you are getting a dose of tannins. While the first example may be a more pleasant experience than the second, the effect in your mouth is the same: the tannins bind to your salivary proteins and make your mouth feel drier. This doesn't hurt us, but insects that feed on tannin-rich leaves may have so many of their salivary and digestive proteins bound up that they aren't able to get much food value from the leaves. Too much of a tannin-rich diet may prevent them from surviving and reproducing. Successful insects look for something a little less astringent.

Other deterrents include skin irritants, like the itchy oils of poison ivy, poison oak, and poison sumac, and phototoxic chemicals found in many members of the parsley family, such as giant hogweed. Farm workers and even grocery shoppers have gotten itchy hands from these light-activated chemicals when handling damaged or diseased celery or parsnips.

POISONS. Of course, these chemicals are the ultimate deterrents. Murder mystery fans will be familiar with strychnine (an alkaloid found in seeds of an Asian tree) and digitalin (a cardiac disruptor found in foxglove leaves). Other less infamous poisons can interfere with growth and reproduction of herbivorous insects. It's not surprising, then, that many of these compounds, like pyrethrins, neem, rotenone, and nicotine, are widely used as pesticides. (But just because these are natural pesticides, it doesn't mean they are safer than synthetic ones.)

ATTRACTION

Plants not only produce secondary compounds to repel other organisms, they also emit chemical signals into the environment to attract them. For example, beneficial microbes—bacteria and fungi that

Sending out an SOS

Like clockwork, the first buds on the rose bushes in our sunny front yard emerge in April, followed by an army of aphids that covers the buds entirely. If I happen to see this, I'll set my hand sprinkler on stun and blast them away, but sometimes I'll forget. When that happens, do my rosebuds get sucked away into lifeless husks? No. In fact, they don't show much damage at all. Eventually our local lacewings and ladybugs stop by for a little green snack.

How do these beneficial predators know where the aphids are? Many plants, ornamentals and vegetables alike, send out very specific gaseous signals when they're under attack. Over time, certain species of predatory and parasitoid insects have learned that these airborne alarms mean lunch. These signals are only emitted during the day, when natural enemies are active. When the herbivorous pests are gone (having either escaped or been eaten), the compounds are no longer produced. To make this phenomenon even more fascinating, plants downwind of the victim may also pick up on the signal and start building chemical defenses against future attack by the herbivore.

When we gardeners indiscriminately spray pesticides for the slightest pest problem, we not only kill the pest, but also the beneficial organisms that could take care of our problem for us at no charge.

provide food, protection, or other benefits—often rely on chemical cues from receptive plants as their invitation to begin a partnership.

POLLINATION. Plants are beguiling in the scents and colors they use to attract birds, bees, and a whole slew of other animal pollinators. Essential oils and other volatile compounds lure pollinators with promises of food or even sex. Your nose might even have helped in pollination as you moved through your garden, taking in all the fragrances. Flower colors are provided by secondary compounds and can be quite specific in their targets. Birds, for instance, see best in the red end of the spectrum— hummingbirds are an obvious example. In contrast, bees prefer flowers at the blue end of the spectrum and can even see into the ultraviolet range. Though we can't see them, many flowers take advantage of bee vision and provide pollination guides leading to the center of the flower that can only be seen under ultraviolet light. If you were a bee, you'd pretty much be seeing giant neon signs reading "Eat here!"

DISPERSAL. After successful pollination, scents and colors are used to attract animals as seed dispersers. The volatile compounds associated with ripe strawberries, peaches, and other summer delights, heated and released by the sun, are irresistible to us and the other fruit eaters we race against. Colors, provided by anthocyanins, carotenoids, and other pigments, are important visual cues that alert fruit lovers to edibility. Whereas most green fruits are hard and bitter, their more mature red, blue, or orange forms are magnets for attention.

GROWTH

ALLELOPATHIC CHEMICALS. Because plants cannot move around to find the resources they need, they can be fierce competitors, exploiting resources and sometimes interfering with their neighbor's ability to do the same. Allelopathic chemicals (meaning killer of others) seep from roots, leaves, fruit, and/or bark of some species, endlessly frustrating gardeners who can't figure out why nothing will grow under their walnut, eucalyptus, or tree-of-heaven. Gardeners have taken advantage of some of these allelopathic chemicals to use as natural herbicides.

PHYTOHORMONES. Some secondary compounds either stimulate known plant growth substances or take this role on for themselves. You'll notice I said "plant growth substances" and may wonder why I'm so stuffy that I can't just say "plant hormones" like other people. The main reason is that it causes confusion. People naturally assume that plant hormones behave like hormones in animals. In reality, they don't. In general, phytohormones can be produced by all plant cells, in contrast to animal hormones that are produced in glands. So, I'll compromise with the term phytohormones, as long as you remember that they are very different from those swarming around in your own body.

Phytohormones can stimulate or inhibit plant activities and control everything from growth to reproduction to death. We'll consider

Vitamin B1

THE PRODUCT
Vitamin B1 (often mixed with fertilizers or plant hormones) is added to new transplants.

THE SUPPOSED BENEFITS
Vitamin B1 stimulates new root development and reduces transplant shock.

HOW PLANTS RESPOND
Plants make their own vitamin B1, so adding it does absolutely nothing for the plant. However, those formulations that include plant hormones such as auxin can help root development. If you are installing picky plants or are working in difficult soils, you could try a rooting hormone. But don't bother with these mixed products, especially those that contain phosphate fertilizer.

LEFT: **Cytokinins produced by insects that lay their eggs within leaf tissue often create galls as nurseries.** RIGHT: **Uncontrolled growth in this conifer has created a witches' broom at the top.**

some specific phytohormones later, but it's worth introducing the main groups.

Auxins are phytohormones that control rooting, stem elongation, and directional growth. Charles Darwin was the first to propose the existence of this group. Gardeners know about auxins because they're used as rooting powders that cuttings are dipped into. As we'll discover later, they're also important in controlling branching. Actively growing shoot tips crank out auxins, which are shuttled downward to other parts of the plant. Auxins are also instrumental in directional growth, causing shoots to grow toward light and roots to grow downward.

Cytokinins are involved in cell division or cytokinesis. Many are manufactured in the roots and transported to shoots, but they're also found in young leaves and fruits, tissues that are rapidly growing. I've referred to them at times as fountain-of-youth phytohormomes, because they can help delay the aging of leaves. Gardeners are familiar with burls, galls, and witches' brooms, which are often caused by foreign cytokinins made by bacteria, fungi, viruses, nematodes, and even insects. The supercharged growth rate of the infected host provides a constant source of nutrition and protection for the invader—a botanical room and board as it were. This relationship can sometimes benefit the plant, too: nitrogen-fixing bacteria create nodules and mycorrhizal fungi develop filamentous hyphae on roots through this same process.

Gibberellins regulate plant height and flower, fruit, and seed development. The first gibberellin was identified in rice plants in Japan suffering from foolish seedling disease. This wonderfully named disease causes rice seedlings to grow so tall that they fall over. It also reduces the affected plant's ability to set seed. Gibberellins can be thought of as the maturing phytohormones—kind of the opposite of cytokinins. As you might suspect, dwarf cultivars of many plants don't synthesize much gibberellin, and adding gibberellins to dwarf cultivars allows them to grow normally. The commercial use of gibberellins has been valuable in increasing fruit size, especially of seedless cultivars like grapes. G is for giant grapes and gibberellins!

Abscisic acid was named for its presumed role in abscission, the deliberate shedding of leaves, fruits, and other plant tissues by programmed cell death. Abscisic acid is particularly active during times of environmental stress, especially any stress associated with lack of available water. In fact, abscisic acid can directly reduce water loss by closing off the leaf pores (stomata), tiny portals through which gases and water vapor travel. Abscisic acid also regulates seed and bud dormancy. Generally, it's considered to be an inhibitory phytohormone, as it tends to slow growth rather than enhance it.

Ethylene is a gas. Really! Put green tomatoes into a bag with a ripe banana, and ethylene released from the banana seems to magically

Overwatered plants often drop their lower leaves due to the release of ethylene from stressed roots.

transform rock-hard tomatoes into soft and juicy delights. Ethylene is also produced by various stresses, so it teams up with abscisic acid in plant responses such as leaf drop, wound sealing, and disease resistance. Like abscisic acid, ethylene tends to be an inhibitory phytohormone associated with programmed tissue death. If you perpetually overwater a container plant, eventually you will see the results of the ethylene gas produced by drowning roots: the lower leaves turn yellow and fall off, leaving only the leaves on top of a bare stem.

Brassinosteroids are a tough-sounding group of phytohormones. They are chemically related to testosterone and other steroids, but you don't have to worry about your marigolds getting muscle-bound. Brassinosteroids are synthesized throughout the plant and are

Biostimulants

THE PRODUCT
Also called metabolic enhancers, biostimulants include humic acids, kelp, probiotics, and other low-nutrient products.

THE SUPPOSED BENEFITS
Biostimulants improve various plant responses including growth, yield, and stress resistance, possibly through stimulation by plant hormones found in the product.

HOW PLANTS RESPOND
Plants make their own phytohormones, and an intact plant is unlikely to benefit from additional hormones applied to the soil or leaves. While there is some value to using hormone dips to root cuttings, there is no consistent positive effect of biostimulants used in any other manner. Proper plant selection, installation, and management practices that ensure development of a healthy root system are more likely to improve growth, yield, and stress resistance than biostimulants. In fact, that's what research has shown.

Leaf stomata consist of pores
bordered by two guard cells.

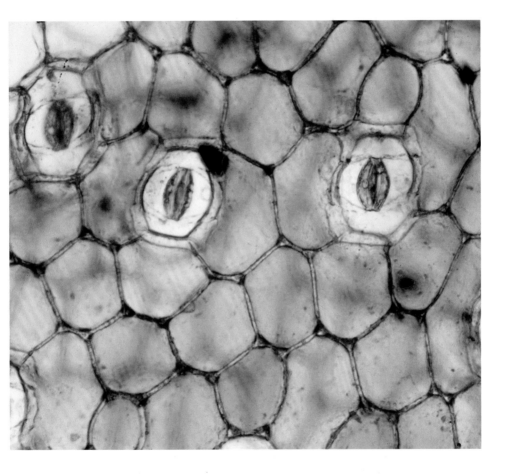

involved in all aspects of growth and development, both stimulatory and inhibitory. They've been used commercially to improve crop yield in plants growing under stressful conditions and can be used like auxins to root cuttings.

Putting It All Together

We've looked at some of the unique qualities of plant cells, which help us understand—and sometimes predict— how plants will respond to their environment. We'll be using this information in later chapters. Let's look at a few nifty plant tissues under the microscope to get us started.

STOMATA

If you were to look at the lower surface of a hydrangea leaf from my garden using a strong hand lens, you would see stomata (also called stomates), pores scattered like tiny pouty lips over the surface. Stomata are the leaves' gas and water regulators. Whereas people have an active respiratory system that brings in the good air and lets out the bad, oxygen and carbon dioxide move passively through these stomata. The pores are also the exit doors for water moving through the plant.

Simple stomata have a pore bounded by two kidney-shaped guard cells, which are attached to one another at their ends. When the guard cells are full of water, their sides push away from each other, opening the pore to the environment. When the leaves wilt, the guard cells are flaccid and the edges seal over the pore, thus reducing water loss.

You'll want to remember how this works later when we look at environmental conditions that cause stomata to open and close and how that affects plants.

APICAL AND LATERAL MERISTEMS

We're back with the microscope, this time looking at clusters of visually unremarkable cells at the tips of roots and shoots. It's at

When mature, these rapidly dividing young root tip cells will have specific functions.

The cell wall is composed of a latticework of cellulose strands. When the crosslinkages are severed, the strands can slip by each other, allowing the plant cell to elongate.

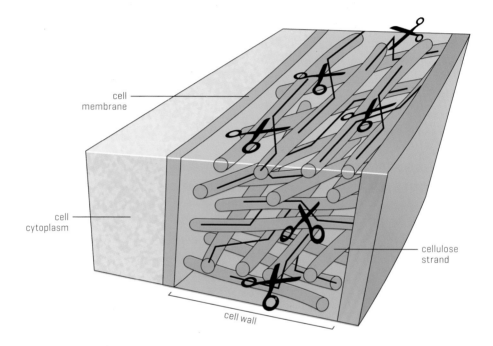

cell membrane

cell cytoplasm

cellulose strand

cell wall

these points, the apical meristems, where cells are rapidly dividing and elongating, allowing roots and shoots to lengthen. Beyond this region, mature plant cells become specialized for water uptake, storage, or some other function. Growing points are also contained in buds tucked into the junctions between leaf petioles and stems; under the right conditions, these buds will develop into branches or flowers. And one group of plants, the grasses, grows from the base rather than the tips.

Apical meristems allow stems and roots to lengthen, whereas

other areas of growth increase the girth of branches, trunks, and roots. These are the lateral meristems, found just under the outer protective layers of bark. These lateral meristems include the vascular cambium, which produces new conducting elements (xylem and phloem), and the cork cambium, which produces bark-related tissues. Lateral meristems are found only in perennial plants that develop girth with age.

"But wait!" the curious gardener will exclaim. "You said that cell walls are rigid. How can they elongate?" We'll start with an easy demonstration to visualize this. Intertwine your fingers together, squeeze them together tightly, and then try to pull them apart. Now relax your fingers and slide them apart without separating them completely. The elongation process in the cell walls is very similar. The cellulose strands in the wall become looser and slipperier during the growth phase, like strands of spaghetti, allowing the cell to lengthen. Once this window of opportunity closes, the strands are locked into place and the cell has formally reached its maximum size.

It's fun to step back from the microscope at this point and consider some analogies. The meristematic cells, undeveloped and actively growing, are like children. As growth slows and they mature, cells take on separate identities and functions, much like teenagers and young adults choose career paths. And of course, the development of girth is something seen in older plants and people!

Understanding where plant growth can and cannot take place will help you understand how plants respond to gardening activities like pruning and staking, as well as how you can do it correctly.

We've had a very quick and selective introduction to plant anatomy and biochemistry. Congratulations on slogging all the way through! You'll find it was well worth your time and patience in subsequent chapters. And I guarantee you will revisit this chapter, both as you continue exploring this book and your own garden.

In any case, I hope you don't feel you've been left in the dark. That's a condition best suited for roots, which may be out of sight, but they certainly shouldn't be out of mind.

The Underground Railroad

TUCKED IN THE CORNER of my backyard is a small opening, covered with a thick layer of wood chips. It's bounded by an arborvitae hedge on two sides, a redbud tree, a dogwood, and several smaller shrubs and groundcovers. Pulling back the moist, decomposing mulch, I can see a mass of fine white threads as well as fibrous roots. As I dig deeper, my trowel thuds against thicker, woodier roots. This complicated network of thick and thin roots, intersected by those fine white threads, runs in all directions like a city roadmap. Yet because it's hidden underground, we often neglect this network—or worse, damage it— through our activities. And those white threads? They're not roots but beneficial mycorrhizal fungi, which we'll talk about later in this chapter.

Fungi and roots create an underground network.

Plant roots serve several functions: they anchor and stabilize the plant, store food, produce growth substances, and take up water and minerals. They represent the beginning of a plant's internal transport system, somewhat similar to our own circulatory system. But rather than being driven by an active pumping heart, transportation in plants is passive and depends on a water gradient that begins in the soil and ends in the atmosphere.

Like the spokes on a wheel, roots radiate outward from the base of the stem, colonizing the soil and tapping pockets of water and nutrients. Some species have rhizomes (underground stems) that grow entwined with the roots. They serve as hidden growing points from which new roots and shoots arise. In fact, some of the clusters of goldenrod, poplar trees, and other common plants are nothing more than clones of a single individual whose roots and rhizomes have taken over a piece of land. We see this in our own gardens with spreading groundcovers and perennials.

How Far Roots Spread

One of the gardener's standard reference points is the dripline, that invisible circle on the ground corresponding to the outermost leaves on a tree or shrub. We mulch, water, fertilize, and protect the soil between the trunk and the dripline, because the roots are there. But, in fact, roots grow beyond this circle. It's estimated that woody plants may have root systems that are at least two to three times the diameter of the dripline. Most of this far-reaching system is made up of fine fibrous roots that are difficult to see when we're digging in the landscape. It's not surprising we don't realize those tiny roots creeping into our annual border are from that maple tree 20 feet away.

Root growth is opportunistic. Fine roots that happen to hit pockets of water, nutrients, or oxygen in the soil will scavenge these resources and push onward. If roots hit a dead end, however, they die back to a larger root. If you were able to watch the root zone under time-lapse photography, you'd see a flurry of activity as tiny roots zoom in every direction, sometimes morphing into tiny starbursts of filaments

when they hit the mother lode. Roots that survive the exploratory phase thicken and harden, eventually becoming permanent storage organs as well as parts of the transport system.

Unlike woody roots, fine roots can be transient. A good example of transient roots can be found in the mulch in my garden that we were exploring. In the wet months of the year, the mulch is always moist and roots find their ways upward to take advantage of the oxygen and nutrients in this upper layer. The aboveground parts of many plants undergo dormancy, but roots grow all year round. As the growing season approaches, temperatures get warmer, rains tend to decrease, and the top of the mulch layer will begin to dry out. The roots that are there will die back to deeper parts of the mulch and soil where it's still moist enough to support growth.

This observation has led to well-intended but erroneous advice about using mulches. You may have been warned that you shouldn't use deep organic mulches because the roots grow into them and eventually

Amending Soil before Planting

THE PRACTICE
Adding organic matter, sand, gravel, or other materials to improve soil fertility, drainage, and/or reduce compaction.

THE SUPPOSED BENEFITS
Plants will establish better with richer, well-drained soil.

HOW PLANTS RESPOND
In limited spaces such as containers or raised beds where the entire soil volume can be amended, plants may establish better depending on the amendment chosen. In landscapes, however, only a fraction of the soil profile is changed by amendment. Where the amendment meets the native soil, you've created a discontinuity that slows down air and water movement. Imagine putting a handkerchief over your nose and mouth. You have to work harder to pull oxygen through it, and the inside of the cloth gets damp from the moisture in your exhalations. It's the same phenomenon in the soil, and likewise plant roots will struggle to cross the barrier.

Although some plants don't really care (heck, some plants grow through concrete), others are very picky about their root habitat and will grow only in the amended area. For smaller plants this might not be an issue, but it will be a problem for woody plants. Instead, use that great organic matter as part of your mulch layer, and leave the soil au naturel.

they'll die. But the transient, opportunistic nature of the fine root system means this is okay—actually, it's more than okay. The plant is able to absorb water, oxygen, and nutrients while conditions are favorable, if only for a short time. These resources are stored and contribute to increased growth and vigor of the entire plant. So mulch away!

How Deeply Roots Grow

When I was growing up, I would often wander through the Douglas fir woods near my home. Once in a while, I'd find a tree that had blown down and I'd scramble onto the weblike mass of exposed roots. Even though I knew, subconsciously, that this was what tree root systems looked like, I still passively accepted the widely held notion that all plants except monocots (such as grasses and orchids), ferns, and mosses had a single taproot that descended far underground, with lateral roots branching off from the sides. Like a Rorschach blot, the symmetry of this appeals to us: a single trunk balanced by a single root.

Our misunderstanding of root systems arises from the fact that dicots have a taproot. But this taproot is transient. It's dominant during the seedling growth phase, when a single root extends from the seed and grows downward. But over time, most taproots are absorbed into a maturing root system that extends laterally as well as downward. For trees and shrubs, the root system may not grow much deeper than a foot or two. As you now know, the lateral spread can extend far past the dripline of the crown.

Why are root systems so shallow relative to their spread? Unlike leaves, roots can't make their own oxygen, so their growth is limited to areas of the soil that are well aerated. Gardeners who make the mistake of planting trees and shrubs too deeply will see these specimens become straggly and listless, until they eventually succumb to the pests and diseases that thrive on injured plants.

Obviously, soil structure and texture will determine how deeply oxygen, and thus roots, will permeate. Compacted soils, which are so common in our urban areas, hold little oxygen. Even seemingly harmless footpaths winding through parks compact soils to the point that

LEFT: **Mature trees and shrubs exploit water, nutrients, and oxygen with broad, shallow root systems.** BELOW: **A long taproot helps seedlings establish quickly.**

they structurally resemble solid rock. The most tenacious weeds have a hard time getting established on these barren soils, much less any desirable species.

In contrast, excessively drained, sandy, or rocky soils are highly aerated and the roots of naturally occurring trees might extend many feet into these unusual soils. The limiting factor for plants in highly aerated soils is water, rather than oxygen.

How Water Moves through Plants

To begin, we need to consider one of the amazing properties of water, specifically the asymmetry of the molecule, which contains two hydrogen atoms and one oxygen atom (H_2O). Imagine your head is

Landscape Fabric
(Geotextile, Weed Barrier)

THE PRODUCT
Landscape fabric is laid on top of soil to keep weeds out while letting water and oxygen in.

THE SUPPOSED BENEFITS
Because the fabric is porous, water and oxygen will pass through to the roots of desirable plants but weeds can't poke through.

HOW PLANTS RESPOND
Unlike the claims right on the packaging, these products do not let water and oxygen through for very long. Those little holes are quickly filled with soil particles, and water puddles on top of the fabric, only slowly dripping through to the parched roots below. Dust and soil blow in along with weed seeds, and within a few months—Voila! Weeds spring up like magic. Likewise, aggressive weeds like bindweed and horsetail slip underneath and pop right through the holes and seams. I guess they didn't read where the package claimed "permanent weed control."

Meanwhile, your tree and shrub roots are desperately seeking water, oxygen, and nutrients. They will creep up to the soil surface, sometimes growing through and on top of the fabric. This causes the fabric to break down even faster. If you try to remove it, you damage your trees and shrubs in the process. Do yourself a favor and don't buy this stuff!

Weeds love landscape fabric.

Because of the slight positive and negative charges at the poles of water molecules, hydrogen bonds (dashed lines) form between them. Hydrogen bonds cause water molecules to cling to each other and to other polar molecules.

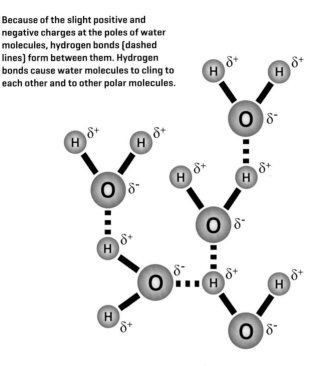

an oxygen atom and your hands are hydrogen atoms, joined by your arms in between (the bonds). Now make a Y with your arms (perhaps performing a molecular version of The Village People's "YMCA"). There is a partial negative charge associated with the oxygen atom and a partial positive charge associated with each of the two hydrogen atoms, and this uneven electrical charge defines water as a polar substance.

We know that opposites attract, and that's exactly what happens with the charged areas of water molecules. Positive regions of polar molecules are attracted to negative regions of others, such that water molecules cling to one another (cohesion) as well as to other polar substances (adhesion). Cohesion allows water to form droplets. If a water droplet is placed on a material made of polar molecules, such as tissue paper, the droplet readily spreads out and adheres to the paper.

But if that same water droplet lands on the waxy surface of a leaf, it remains in droplet form. That's because waxes are made of nonpolar molecules, so there's no electric charge to attract the water.

Tissue paper and plant cell walls are both made of cellulose, a polar material closely related to starch. The water uptake vessels in plants, collectively called xylem, are constructed of nonliving cells that essentially form hollow tubes of cellulose. Water adheres to the sides of the xylem tubes, creating an unbroken stream of water through the stem of the plant. You can think of them as giant straws, sucking the water from the roots to the leaves.

Finally, we need to understand how water can move from one place to another without the addition of energy. The basic rule is that water moves toward areas that contain less water. Many gardeners know what happens when they put salt on a slug. The slug shrivels up and dies as the water in its body moves outside toward the salt: the area of the body with salt on it has less water than the slug has on the inside. This passive diffusion works in plants, too. Moist soil contains lots of water, which diffuses into roots, where the presence of sugars, amino acids, and other dissolved substances means that water content is relatively lower compared to the soil. Leaves have even less water, because they have even more dissolved sugar. So water moves from the roots, through the stems, and into the leaves. The final pull comes from the atmosphere, where water concentrations can be quite low, especially when it's sunny or windy. Water is pulled through the stomata, tiny pores in the leaf surface, into the atmosphere, dragging chains of water molecules behind in a process called transpiration. It's estimated that 97 percent of the water taken up by the roots is lost to the atmosphere.

Why don't plants conserve water more efficiently? Part of the answer is that transpiration is more than just a process for moving water through the plant. Minerals, nutrients, and other dissolved substances can all hitch a ride on the Transpiration Express. These tagalongs are especially noticeable in the spring, when the roots release materials they have stored all winter, as anyone who's ever tapped a maple tree to make sugar or syrup will know. During the

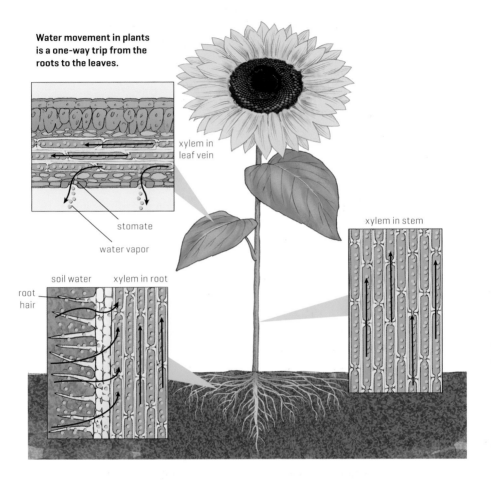

Water movement in plants is a one-way trip from the roots to the leaves.

xylem in leaf vein

stomate

water vapor

xylem in stem

soil water

xylem in root

root hair

xylem in root

hot summer months, water loss from the leaves lowers the leaf temperature through evaporative cooling, keeping the photosynthetic machinery in a healthy state.

Nevertheless, the apparent waste of water still puzzles both gardeners and scientists, who often wonder why plants haven't developed a more efficient way to take up carbon dioxide for photosynthesis. For it is through the stomata that carbon dioxide enters the leaf, at the same time that water exits. Cacti and other succulent plants of arid

Sheet Mulches

THE PRACTICE

Placing newspaper or cardboard over weeds or bare soil. Sometimes compost or wood chips are put on top, often in many layers to create a sort of mulch lasagna.

THE SUPPOSED BENEFITS

Sheet mulches kill existing weeds and prevent weed seeds from germinating. They're better than landscape fabrics because they break down and contribute to soil health.

HOW PLANTS RESPOND

It's true that newspaper and cardboard sheet mulching is better than landscape fabric, but that's not much of a compliment. Any sheet mulch creates a barrier to oxygen and water transfer between the soil and the environment. Roots in such a situation will grow closer to the surface of the soil to grab any oxygen that permeates through the sheets. Likewise, worms and other soil organisms move closer to the surface as the soil becomes more oxygen depleted, and gardeners mistakenly attribute their increased visibility to the attractiveness of the sheet much. When paper-based sheet mulches dry out, irrigation and rain water run off them, creating arid conditions underneath.

A virtual desert exists underneath dried-out cardboard mulch.

Paper-based sheet mulches do break down, but there is no scientific rationale behind their use. Chunky, coarse organic mulches are more effective in reducing weeds and create a more hospitable environment for roots and soil critters alike.

A visible root flare is crucial
to tree longevity.

climates have evolved water conservations strategies, such as closing the stomata during the day and only opening them at night. This
adaptation affects the plants' photosynthetic ability (as we'll explore
more in chapter 4).

How to Tell When Roots Aren't Happy

Plants can survive a long time with root problems. The key word here is
survive, they certainly don't thrive. Observant gardeners might notice
that a recently planted Japanese maple just doesn't seem to get much

LEFT: The burlap and twine on balled-and-burlapped trees protect them in the nursery, but they interfere with root establishment when the trees are planted. RIGHT: The circling roots of container plants will continue their orbit unless straightened out upon planting. (New roots, however, will grow outward.)

taller. Or maybe the new leaves on a favorite rhododendron seem to be getting smaller every year. These are signs that the root system isn't supplying enough water and/or nutrients to support vigorous growth aboveground. Figuring out why is tricky, especially without exhuming the roots. There are some key symptoms, however, that gardeners should look for. Leaves may be smaller or sparser than normal, or they may have scorched tips and margins in the summer, all because they aren't being supplied with enough water. Assuming there is enough water supplied to the soil, what else might be happening underground?

HOW LOW DOES IT GO?

Let's start with the recently planted Japanese maple—or any other young tree—that stubbornly stays about the same size year after year. Can you see the root flare, the part of the tree where the trunk morphs into roots? This flare should be at the soil surface. If you can't see it, the tree is planted too deeply. Although some tough tree and shrub species don't mind this treatment, others have roots that will struggle for oxygen. Roots without enough oxygen aren't very efficient at taking up either water or nutrients, effectively starving the trunk, branches, and leaves. You can commute the death sentence on these trees by digging them up and replanting them correctly. Do this within the first year after planting and you'll probably be successful, but the longer you wait the more you reduce your chances of success.

Even properly planted trees and shrubs are in danger if their roots are suddenly buried due to changes in grade. Maybe you've put an addition on your house, widened a driveway, or done some other work where the soil level has been increased. Adding several inches of soil on top of existing fine root systems is literally burying them alive. Unlike leaves, roots don't make their own oxygen and they depend on pockets of air trapped in the soil. When the soil suddenly becomes much deeper, it's more difficult for oxygen to seep through from above and the roots suffocate. Mature, well-established trees might be able to survive this treatment, but younger plants will benefit from being exhumed and replanted at grade.

SOLITARY CONFINEMENT

Now let's consider that rhododendron whose new leaves are obviously smaller than those from the year before. You can see the flare of the roots, so you know it's not buried too deeply. I call the next diagnostic tool the wiggle test. Gently grasp the plant near the base and wiggle it. Can you see soil movement around the root ball? Or can you feel movement through the trunk? Your fingers are exquisitely adapted to sense differences in resistance. With this rhododendron, you can see the root ball rock, so you know that the roots have not established in the soil. Now you have to figure out why.

Since the roots are obviously not exploring the surrounding soil, there's no reason not to dig up the rhododendron. What you're likely

Bare-Rooting Trees and Shrubs for Planting

THE PRACTICE
Removing all foreign materials—containers, wire baskets, burlap, twine, and soil—from woody plant roots before planting.

THE SUPPOSED BENEFITS
By removing as much of the foreign material as possible, roots will be in direct contact with the native soil when planted.

HOW PLANTS RESPOND
This can be a stressful process for *you* as well as the tree or shrub. We've been taught for so long to leave the root ball intact that it's very difficult to look at the issue objectively. But research has shown bare-rooting provides better long-term establishment and survival than leaving the root ball untouched.

Left intact, the clay or soilless media that is nothing like the site's native soil will create barriers to water and air movement to the roots. By removing this material, plant roots are immediately able to establish in the native soil without passing through multiple barriers.

Poor root systems can be corrected before planting by removing circling, kinked, or otherwise deformed roots. The root system should look like spokes on a wheel. Also, pruning stimulates roots to grow once planted.

Removing all foreign materials will allow you to unearth the root flare. This structure needs to be planted at grade, not underground.

Finally, the transplanted tree will not need to be staked for more than a month or two, if at all, when planted in this manner.

This is a revolutionary approach and one that does not sit well with many home gardeners or nursery and landscape professionals. However, it is supported by research showing better long-term establishment and survival than conventional methods.

Washing the clay from roots of large and small trees will improve their establishment when planted.

to find is a root ball surrounded in burlap or possibly shaped like the container it was originally planted in. Plants whose roots are isolated from the native soil have a difficult time establishing, and the best thing to do is release them from their solitary confinement. If it's a balled-and-burlapped root ball, you'll want to remove the twine, burlap, and as much of the clay as you can. Cylinder-shaped root balls from containers need to be cut and spread to correct the circling you'll see. This might seem like plant abuse, but it's the only way to deal with incorrigible root systems. The cool thing about roots is that pruning stimulates new root formation. So take the tough love approach with these anti-establishment trees and shrubs. They'll thank you later!

Fungal Alliances

I've compared root systems to an underground railroad, but imagine upgrading that freight train to high speed rail. That's what a certain group of fungi can do for your landscape and garden plants underground. We saw them in my backyard at the beginning of this chapter: those white threads too long and slender to be roots, yet cozily intertwined with the root system. These are the hyphae of fungi, and their associations with plants are collectively called mycorrhizae (meaning fungus roots). Before you grab the fungicide, let's get to know these plant partners.

Mycorrhizae are ancient and beneficial associations that arose hundreds of millions of years ago, when plants first extended roots into soil. When first discovered in the late 1800s, mycorrhizal relationships were thought to be unusual oddities. We now know that they are the rule, rather than the exception, especially in woody plants. Mycorrhizal relationships are mutualistic in that both partners receive a significant benefit in exchange for sharing resources: you scratch my back, and I'll scratch yours. Plants transfer carbohydrates and B vitamins through the hyphae to the fungi, which are not photosynthetic and can't generate their own food. In return, the fungi extensively colonize the root surfaces and enhance the plant's uptake of water and mineral nutrients.

Mycologists, scientists who study fungi, have divided mycorrhizal fungi into two categories depending on how cozy the relationship is. Those whose root-like hyphae surround and occasionally penetrate root tissues are ectomycorrhizae (*ecto* means outside), and those whose hyphae always enter the root cells are endomycorrhizae (*endo* means inside). Ectomycorrhizal fungi partner up with many woody plant species, forming an extensive hyphal network throughout mulch and topsoil layers; these are the networks we saw in my backyard soil. In contrast, endomycorrhizae are found in hundreds of plant families. You might see these delicate structures associated with the roots of your annual flowers or vegetables.

AN INFECTION YOUR PLANTS WILL LOVE

Healthy soils contain vast repositories of mycorrhizal spores in the coarse organic matter near the soil surface, where they germinate under moist, aerated conditions. As the hair-like hyphae emerge from the spores, rain and irrigation water create channels down through the soil toward growing plant root tips. Receptive roots release chemical cues that allow hyphae to penetrate the cell walls and create chemical passageways between the two partners. Multiple infection points and hyphal branching create a cottony sheath around the roots that extends far into the surrounding soil. Like microscopic miners, mycorrhizae discover and extract soil water and nutrients from otherwise inaccessible pockets.

The impact of mycorrhizal colonization goes far beyond an individual plant. Most plants are colonized by a variety of mycorrhizal fungi, and most of these fungi have multiple hosts. Mycorrhizae can link roots of different species, transferring nutrients to the plants with highest demand. At the same time, the dense network of fine hyphae increases soil aggregates and improves soil stability, while enhancing organic matter decomposition and acidifying the root zone. The resulting network is a virtual fungal freeway of nutrient and water acquisition and transfer.

As if extra water and nutrients weren't enough of a benefit for plants, their fungal partners have some bonus gifts for joining the network. Mycorrhizal plants are more resistant to environmental stresses, such as drought and salinity, because their ability to find and extract soil water is improved. The dense fungal network created by mycorrhizal fungi limits the available space for colonization by pathogens or attack by nematodes, meaning that plants are less prone to disease and root pests.

But for this vital relationship to exist, plant roots need to be receptive to inoculation. The best way to ensure this is to avoid overwatering and excessive fertilization. Adding too much fertilizer, especially those containing phosphate, is by far the most damaging garden practice in terms of mycorrhizal health. Composted manure and many soilless potting mixes contain high levels of phosphate and other nutrients. With a plethora of nutrients, plants are less dependent on mycorrhizal connections. Mycorrhizal fungi retreat into the shadows, remaining inactive until more hospitable soil conditions return.

Not surprisingly, this mutually beneficial association between mycorrhizal fungi and plants has been marketed in the form of products that we're told will improve soil health and plant establishment in gardens and landscapes. But scientific studies continue to show that these products have no value. We know that healthy soil will naturally contain a smorgasbord of beneficial microbes, including mycorrhizal fungi. And if a soil is in such bad shape that native mycorrhizal species don't survive, adding packaged spores isn't going to help.

Don't Derail the Train

As you can probably tell by now, healthy vigorous root systems and their mycorrhizal partners will reward you with a garden full of lush, attractive plants that require few fertilizers or pesticides. Even though you can't see this busy underground railroad, you need to constantly be aware of its presence and tread lightly. Activities associated with construction are extremely damaging to soil structure. Topsoil is removed, taking with it the beneficial microbes and all of the organic material. Adding insult to injury is the compaction caused by heavy equipment, which creates oxygen-depleted soils about as hospitable as cement. Of course, mycorrhizal colonization and plant communities will eventually recover, but unnecessary soil disruption should be avoided. This also means avoiding gardening activities that compact the soil, or worse, grind it into a lifeless powder.

Yes, I'm talking about rototilling. I know a lot of you enjoy your power tools and handling a rototiller is almost as fun as riding a bucking bronco. But as far as life in the soil is concerned, this is the equivalent of an underground tsunami. Rototilling destroys natural soil structure along with any plant roots and hapless animals in the path of destruction. Soils are more than just a medium for growing veggies: they are complex ecosystems containing beneficial bacteria, fungi, insects, nematodes, earthworms, and many other denizens. Well-structured soils, along with their natural living communities of organisms, benefit plant roots and enhance their establishment. Roots damaged by rototilling require energy and resources to repair, and when their protective outer tissues are torn they are exposed to diseases and pests.

In this case, less is more in terms of soil and plant health. Instead of revving up the rototiller, try using hand tools and digging only where you intend to place seeds or plants. When you're getting rid of sod or other unwanted plants, cover them with a deep layer of arborist wood chips and let them die a sunless, hidden death. You'll preserve soil integrity and reduce fossil fuel consumption as well. Add some mulch to protect any exposed soil, water well, and your impact on this complex ecosystem will be minimal.

Looking at the corner of my backyard again, it's with new respect that we consider the landscape underground. Roots and their fungal helpers mine the soil for water and minerals, efficiently transporting this nutritious liquid to the tops of my redbud and the arborvitae hedge without a pump. Which gets us to wondering, what are all those nutrients used for?

What's
Essential

WE GARDENERS TEND TO FUSS about our garden plants' nutrition. We worry whether we're providing the right form of phosphorus, if organic forms of nitrogen are better than inorganic ones, and if we need to add chelated iron or kelp or humic acids or mycorrhizal fungi. Whew.

The fertilizer sections of nursery and garden centers that have exploded in the last decade are beginning to look like nutritional supplement stores. How in the world can you find out what exactly you should be buying? In other words, how can you determine what's essential for your garden plants and what's just a clever marketing strategy?

Strictly speaking, the essential elements are those that plants require to complete their life cycle. This chemical collection is constantly changing as we learn more about plant biochemistry. When I took my first class in plant physiology, I think there were fifteen elements labeled as essential. Now we recognize at least nineteen of them for most plants, and some specialized plants require still others.

The way researchers determine whether an element is essential to plant growth is through a nutrient exclusion experiment. Seedlings are grown hydroponically in solutions that contain all the known nutrients except for the one in question. Researchers can discover whether the plants reach maturity and reproduce (in other words, complete the life cycle), as well as observe what deficiency symptoms look like. As you might imagine, it becomes more and more difficult to run these experiments with nutrients that are required by plants in increasingly smaller amounts.

Although it's easy to do nutrient exclusion experiments and see how a particular species responds to a missing element, it's extraordinarily difficult to do this in the garden. Toxic levels of one nutrient can mimic deficiencies of another, and viruses and other environmental factors can cause symptoms similar to nutrient deficiencies.

Yellowish leaves on this bean plant might indicate a nitrogen deficiency.

Instead, gardeners should rely on soil tests to discover which nutrients are deficient and, more importantly, which are in excess.

Mineral nutrients were historically divided into macro- and micronutrients, depending on how common they are in plant tissues relative to one another. But again, as our understanding of plants improves, we discover that species can vary dramatically in their mineral nutrient content, so it's more useful to look at what function these nutrients play, rather than worrying about how much of each a particular plant might contain.

The Big Three: Carbon, Hydrogen, and Oxygen

Your garden plants, you, and all other life on Earth are carbon-based life forms. Carbon forms the scaffolding from which other elements are hung, in particular hydrogen and oxygen. These three elements together are the principal building blocks of organic compounds, which include carbohydrates, proteins, fats, and nucleic acids (DNA and RNA). (In a chemistry context, the word *organic* relates to the molecular makeup of a compound. It has nothing to do with the popular notion of equating organic with natural.) Carbon, hydrogen, and oxygen are obtained from the atmosphere and water. Therefore, they aren't considered to be mineral nutrients, which come primarily from the soil.

The Other Three: Nitrogen, Sulfur, and Phosphorus

Nitrogen is an intrinsic part of amino acids, which means it's needed for every protein and enzyme a plant produces. It's also part of the chlorophyll molecule, is embedded in DNA, and is a major component of alkaloids, a group of protective secondary compounds. Soil microbes need nitrogen, too, so this element can become deficient in an actively growing garden as everyone competes for their fair share. If you notice the older leaves of your plants turning yellow during the

Root nodules contain bacteria that convert atmospheric nitrogen to a form that can be used by plants.

peak of the growing season, it might mean your soil nitrogen levels are low. Plants scavenge nitrogen from their older leaves and use it to support new leaf growth, so the old leaves become chlorotic.

Some common garden plants, like peas and beans, avoid this situation altogether by producing their own nitrogen. These plants, which include all of the legumes and members of some other plant families, are nitrogen-fixers, meaning they strip nitrogen from the air and transform it into a solid form. Nitrogen-fixing plants rely on specialized bacteria to transform nitrogen gas into a solid form. The plants are landlords: they provide room and board for their bacterial tenants, who take up residence in the roots in customized nodules. Sugars and other necessities are routed through these nodules, so the bacteria are well fed as their nitrogenase enzymes crank out nitrogen compounds for the plant to use.

A fascinating part of this symbiotic relationship between plant and microbe is that nitrogenase requires a low-oxygen environment to work its enzymatic magic. The plant keeps its root nodules' oxygen levels low by binding oxygen to a specialized protein called leghemoglobin. It's a delicate balancing act, because the bacteria require oxygen for respiration: the levels of oxygen have to be low enough for nitrogenase to function, but high enough so the bacteria don't die. Like the hemoglobin in your own blood, leghemoglobin is bright red when it binds oxygen, surprising many a gardener who has cut one of these nodules open.

Like the relationships between bacteria and ancestral plant cells that led to the development of mitochondria and chloroplasts, nitrogen-fixing bacteria in plant roots might be in the midst of a similar evolutionary process. Right now plants have to become inoculated with the nitrogen-fixing bacteria, but in time the bacterial DNA could become permanently entangled in the plant cells' reproductive process. At that point we'd have a "nitroplast," which, like chloroplasts and mitochondria, would have completely different DNA than the nucleus.

In some environments, nitrogen deficiencies are chronic and soils might not be hospitable to nitrogen-fixing species. These include some wetland soils as well as thin soils over bare rock. In such environments, plants have evolved another way to obtain nitrogen: by trapping and consuming insects and even larger prey. Carnivorous plants can also take up nitrogen from the soil when it's available, but their traps provide a backup pantry just in case. Most of us don't live and garden in areas with such nitrogen-poor soil, but even if we do it's pretty easy to fix the problem with moderate additions of alfalfa meal or some other nitrogen-rich fertilizer.

Sulfur is also required for two amino acids, so it's just as important for proteins and enzymes as nitrogen, though sulfur is not needed in as large a quantity. Because most soils contain sufficient sulfur, deficiencies are rare. When sulfur is deficient, however, the chlorotic yellowing can be distinguished from nitrogen deficiency because it occurs in the younger rather than older leaves. Once sulfur is in the plant it's not very mobile, so it can't be scavenged and reused in young leaves like nitrogen can.

Phosphorus is another building block needed for constructing membranes, energy-containing compounds, and genetic material. It's unusual to find phosphorus deficiency in most home gardens and landscapes, and deficiency symptoms are not as clear-cut as they are for nitrogen and sulfur. Unfortunately, common gardening lore warns that red leaves are a sign of phosphorus deficiency, but red foliage can be caused by many factors, not just nutrient deficiency. In fact, many well-tended gardens often have too much phosphorus due to the overuse of so-called transplant fertilizers including bone meal, guano, chicken manure, and superphosphate. We'll take a closer look

Sugar

THE PRODUCT
Table sugar is sprinkled on soil as an herbicide.

THE SUPPOSED BENEFITS
Sugar stimulates bacterial growth, which ties up nitrogen and prevents weed seed germination and growth.

HOW PLANTS RESPOND
Well, yes, this can work, but it's nonselective. In other words, nitrogen deficiency is going to zap all of your plants, not just the weeds. Weeds are called weeds for a reason: they are aggressive and persistent. If a weed can't tolerate your soil, your annuals and vegetables certainly won't be able to either. There are better ways to control weeds, and you can save the sugar for more delectable purposes.

at the unexpected effects of phosphate toxicity later in this chapter. But before you buy another box of bone meal, I want you to solemnly swear never to add phosphate to your gardens unless a soil test tells you to do so.

Cell Wall Strengtheners: Boron, Calcium, and Silicon

Boron, calcium, and silicon all add structural stability to plant cell walls. You can consider them to be the plant cell's skeleton. Boron has been described as a cellular stapler needed to hold other chemicals together and stabilize cell walls. It might even function as a natural insecticide, because boric acid is toxic to many insects. Calcium not only strengthens cell walls but is important in cell membranes too; it's both a structural component and used for communication between cells.

Every gardener, at some point, will have become intimately familiar

Gypsum

THE PRODUCT
Gypsum is calcium sulfate, an inorganic fertilizer.

THE SUPPOSED BENEFITS
Gypsum improves plant health by improving soil tilth.

HOW PLANTS RESPOND
Once again, adding any nutrient to soil without knowing if it's deficient can create mineral imbalances in the soil with negative effects on plants. The overuse of gypsum can cause soil deficiencies of other important plant nutrients, including iron, magnesium, manganese, phosphorus, and zinc, and it may negatively affect inoculation of roots by mycorrhizal fungi.

Boy, that soil test is looking better all the time, isn't it?

with the importance of silicon in certain plants. This is the element that glass is made of, and if you've ever gotten a grass cut, bingo! That's the silicon home protection system. Pioneers made use of another silicon-rich plant in the pre–Brillo pad era. Horsetails, or scouring rushes, are natural pot-scrubbers because of the abrasiveness provided by silicon.

In home gardens and landscapes, none of these three elements is usually limited. Because these minerals are associated with the cell's skeleton, their absence usually causes young leaves and fruit to look lumpy, bumpy, or otherwise deformed. A soil test is the only way to know for sure if one of these elements is missing.

Chemical Jugglers: Copper, Iron, Magnesium, Manganese, Molybdenum, Nickel, and Zinc

These nutrients are crucial to many enzyme activities, because they shuttle electrons between the chemicals in reactions. These metallic elements easily change back and forth from one charged state to another as a result of accepting and donating electrons, which are negatively charged. This may seem an esoteric bit of information, but many heavy metals in the soil are toxic to plants and to you.

The metallic nutrients are sometimes difficult to dissolve. You can put an iron nail in your garden soil and watch it slowly turn to rust, but this isn't a very efficient way of providing nutrients to roots. Instead, iron is usually available in a chelated form. You might see containers of Fe-EDTA, which is shorthand for iron chelate. Chelate comes from the Greek word for claw, and this is exactly what chelating compounds do: they grab onto individual iron atoms and make them soluble in water. Once the iron has been transported into the appropriate plant tissue, the plant's own chelating compounds trap the iron atoms so they stay soluble and functional.

Deficiencies of some of these metallic elements, such as iron, magnesium, manganese, and molybdenum, create an oddly artistic pattern of leaf yellowing. The veins remain green and the tissue between

Yellow leaves with distinctly
green veins create a pattern
called interveinal chlorosis.

the veins turns yellow, a pattern called interveinal chlorosis. This is a common symptom in many landscape trees and shrubs. Acid-loving plants, like rhododendrons, show this clearly when they're planted in alkaline soils or if they're near newly poured concrete sidewalks or foundations. The lime from the concrete leaches into the soil, which raises the pH and prevents iron uptake by acid-loving species.

Much of the time, however, pH has nothing to do with interveinal chlorosis, and soil tests show plenty of available iron. It turns out that this type of leaf chlorosis is not caused by a soil nutrient deficiency, but rather by excess phosphate, which we'll discuss later in this chapter.

Water Managers: Potassium, Chlorine, and Sodium

These three minerals move freely throughout the plant, managing water movement between the cells and throughout the plant. Potassium is most important in this regard. It's partially responsible for opening and closing the stomata, which are the exit portals for water

Epsom salts

THE PRODUCT
Epsom salts are another name for magnesium sulfate, an inorganic fertilizer.

THE SUPPOSED BENEFITS
Epsom salts increase seed germination, improve nutrient uptake, and enhance overall growth.

HOW PLANTS RESPOND
Epsom salts are often used to treat magnesium deficiency in fruits, vegetables, and timber species. Magnesium deficiency commonly occurs in soils under intensive agricultural production, not your backyard vegetable garden. Most nonagricultural soils contain plenty of magnesium—sometimes too much—and adding more just makes matters worse. Excessive application of other nutrients, like potassium, can interfere with a plant's ability to take up magnesium, making it appear that the soil has a magnesium deficiency when the problem is actually potassium toxicity. Test your soil before you try to diagnose and treat a nutrient problem.

moving through the plant. This mineral also assists with water transport across membranes, which can cause cells and entire tissues to become turgid or flaccid, depending on which way the water is moving. Fortunately, potassium is rarely deficient in home gardens and landscapes. Chlorine is of secondary importance in regulating water movement, but is required in minute quantities; soil deficiencies are highly unlikely.

In some species, such as cacti and succulents, sodium can take the place of potassium. The two elements are similar to one another in size and charge, so it's a convenient substitution for plants that often live in very salty soils. Plants adapted to salty soils are also found in coastal areas, where salt water makes a regular appearance. It's possible that sodium deficiencies could occur in these and other salt-loving species when grown outside their native habitats and away from their normal soils.

Mineral Specialists: Cobalt, Selenium, and Aluminum

A handful of elements are required for growth by only a select group of plants. Probably the most common of these is cobalt. Members of the pea family and some other species can house nitrogen-fixing bacteria in their roots, and cobalt is required for nitrogen fixation. Whether the cobalt is required by the plant itself or its microbial guests isn't clear, but in the absence of cobalt nitrogen-fixing species don't complete their life cycle.

Selenium is another unusual nutrient, seemingly required only by plants like milk-vetch (*Astragalus* species) that colonize selenium-rich soil. In fact, these species are considered to be indicator plants, acting like green arrows pointing to high levels of selenium in the soil. The selenium that accumulates in their tissues is toxic to cattle and other grazing herbivores, explaining the origin of another common name for some species of *Astragalus*: locoweed. Selenium obviously protects accumulator plants from being eaten, but other possible functions aren't yet known.

Many gardeners use aluminum tags to identify their plants. Some of these plants use aluminum themselves, but in a very different way. Hydrangeas are probably the best known garden-variety aluminum accumulator. The gorgeous blue hues some of them sport are due to aluminum in their tissues. Like selenium, aluminum is toxic to animals, and perhaps this is the reason we don't see a lot of insect damage on hydrangeas. People unfortunate enough to experiment with hydrangea tea suffer the same type of neurological poisoning that you'd get from deadly nightshade. Occasionally, younger hydrangea leaves and stems can fall victim to deer and some insects, probably because they haven't accumulated defensive chemicals found in mature tissues.

Future research will undoubtedly uncover other minerals required

The variety of hydrangea colors is
partially due to their aluminum content.

by specialized plants for their survival. But there are some elements
that may never reach the essential nutrient standard because they are
essentially toxic to plants, to animals, and to you.

Heavy Metal

I'm just throwing that title out there so you can figure out if you're a
music freak or chemistry geek. If your first thoughts were of Led Zeppelin or Metallica, my teenage son thinks there's hope for you. But if
you instead visualized lead, arsenic, or mercury, then read on!

Some of the essential elements we earlier labeled as chemical jugglers are heavy metals: iron, copper, manganese, and zinc are four of them. They all transfer electrons by alternatively accepting them from one chemical and donating them to another. There's another group of heavy metals, however, that aren't as willing to pass electrons back and forth. Instead, they effectively shut down the enzyme system that's driving a chemical reaction. These are toxic heavy metals, including lead, mercury, arsenic, and cadmium, which can become an unwelcome addition to your plants' tissues, as well as to whatever eats those plants, including you.

Heavy metals are everywhere. They are naturally occurring elements that happen to have bad effects on living systems. Because they are elements, they don't break down. This is an important distinction to make, because other types of contaminants do break down and become less harmful over time. If your garden or landscape soil contains toxic heavy metals, they are there for good unless you have the soil removed. This is unfortunately a topic that causes gardeners distress, but understanding where metals come from and what you can do about them helps.

FOREWARNED IS FOREARMED

Unless you live in the middle of nowhere, it's likely that your soil will contain heavy metal pollutants of some sort. Lead is by far the most common, as our reliance on lead-based paints and gasoline has deposited plenty of this element into urban and residential soils. Sadly, arsenic is also common, especially in land historically used for agriculture in the early part of the last century. It turns out that arsenic is a great pesticide, and it was used liberally in treating orchard crops and other agricultural fields. Until a few years ago, pressure-treated lumber was laden with arsenic and chromium, both of which slowly leach out of the wood and into the surrounding soil or water. Smelting operations, like the one in my hometown of Tacoma, Washington, dumped tons of arsenic and other heavy metals into the atmosphere, where they might settle out in soils miles away from the smelter. Tires and mulches made from recycled rubber break down to release a constant supply of zinc, cadmium, chromium, and selenium. Even unregulated topsoils and composts can carry heavy metal contaminants; it's always best to purchase soils and compost that have been tested for metal content and certified.

Once in the soil, where do the metals go? Alkaline and clay soils tend to hold them pretty tightly, whereas acid and sandy soils are more likely to release them for root uptake. Plants vary widely in their uptake of metals, depending on the species, life stage, and tissue of interest. The bottom line is that it's impossible to predict with any accuracy what plants are going to take up what metals into which tissues.

That being said, there are some general observations that can be helpful. Roots and leaves are the tissues most likely to accumulate a given heavy metal. So stem vegetables, such as celery, leeks, and rhubarb, are probably pretty safe. Botanical fruits, meaning anything containing seeds, tend to be safe from metal accumulation. This makes sense, because the fruit serves to attract fruit-eaters, who deposit the seeds elsewhere. The same goes for floral products like nectar and pollen. You're not likely to eat much of these, but beneficial insect pollinators are. From a plant's perspective, killing off bees and butterflies is not a great way to reward one's reproductive matchmakers (or produce offspring, for that matter).

Obviously, you should have your soil tested if you are at all concerned about possible heavy metal contamination. Even some of the essential heavy metals, like zinc, can become toxic to plants if they're too concentrated in the soil. Soil testing is the best way to sort out what minerals your soil holds so that you can take appropriate action. And what if those heavy metals do show up in a soil test? Let's look at some constructive ideas.

First, let's address a worst case scenario: your soil has unsafe levels of arsenic from pesticide usage almost a century ago. The fruit orchards razed in the 1940s to build suburbs may be a distant memory, but the arsenic remains an invisible hazard until your soil test brings it to light. You probably will make the decision not to grow vegetables in this soil, though you can still plant turf, trees, shrubs, and other ornamentals instead. If soil removal is not feasible, you should consider building some raised beds. These beds can be constructed from natural wood, plastic timbers, the new version of pressure treated lumber (which contains no arsenic or chromium), bricks, stones, or concrete blocks. Or you can try container gardening, which has been a staple of apartment and condominium living for years.

Once you have your new vegetable area designated, you'll want to purchase certified clean topsoil and compost. Your nursery or garden center will probably carry both. You can also make your own compost, as long as you know that your feedstock is free of contaminants. Avoid planting near roadways; the lead from years of leaded gasoline lurks in road dust and can easily be blown into your vegetables. Finally, use only tapwater or water gathered in rain barrels for irrigation of edible plants; gray water can contain unwanted contaminants.

Hopefully you'll discover that the heavy metals in your soils are at baseline levels—remember, they are naturally occurring elements— and that it's perfectly safe to grow whatever you like. Now you can use your soil test to determine which, if any, nutrients need to be added through fertilizers. You may be pleasantly surprised that you don't need to add much of anything.

Soil Testing: A Wellness Checkup for Your Garden

Let's assume you're feeling run down and suspect you might have a mineral deficiency of some sort. Would you run to your nearest supplement store, buy one of everything, and chug them all down? Of course not! You would probably go to your doctor first and have some blood work or other lab testing done to find out what might be missing. Your garden soils and plants deserve the same treatment.Before you spend a lot of money on fertilizers and other remedies, you need to find out what your soil already contains and what it might need. Otherwise, you could be creating nutrient toxicities by adding too much of some minerals. And it doesn't matter whether the nutrients come from organic or inorganic sources: too much is too much.

It's easy to take soil samples and send them off for analysis. There are many private soil testing labs available, but I prefer university labs. The prices are reasonable, and there is no hawking of soil remedies that you might find with some less savory private labs. Regardless of which lab you choose, you'll need to follow their directions carefully in sampling the soil. You should take samples from several places in your garden area, avoiding the inclusion of any mulch, and then mix them all together before taking a final sample from the mixture. You'll also need to identify what kinds of plants you're interested in growing in any particular area, such as turf, vegetables, annual flowers, or trees and shrubs. If you are concerned about heavy metal contamination, you'll need to ask for these specific tests to be done. They're usually not part of a basic soil analysis procedure.

You may be tempted to use one of the little home testing kits sold at nurseries and garden centers. My advice is don't waste your money. Though these kits may accurately measure pH, they are not sophisticated enough to tell you much more about nutrient content. Just like the medical labs you rely on for blood analysis and other tests, soil testing labs provide the most accurate results. Most importantly, they will provide information explaining what, if anything, you need to do to improve nutrition or address contamination.

Fertilizer Fun Facts

Gardeners are most familiar with fertilizers based on three elements: nitrogen, phosphorus, and potassium. It's the NPK formula on nearly all fertilizer packages. An NPK ratio of 10-10-10 means that the fertilizer contains 10 percent of each of these elements by weight.

With the emphasis on NPK fertilizers, you might well assume that these three nutrients are most commonly deficient in your garden. Nothing could be further from the truth. In fact, only nitrogen tends to run low in garden soils, and then usually only in the summer when everything is growing like gangbusters. At that point, it's easy to give a little dollop of a nitrogen-rich resource if your annual plants look like they're slowing down. (The growth of trees, shrubs, and other perennials tends to slow naturally in the summer anyway, so they aren't good indicators.)

The NPK label on this fertilizer bag shows
that it contains 9 percent nitrogen, 1 percent
phosphorus, and 0 percent potassium by weight.

So why the preponderance of NPK fertilizers in garden centers and
nurseries? It's a holdover from our agricultural past. When farmers
grow crops, they are deliberately creating an artificial system to max-
imize production of the crop over a short period with lots of water
and fertilizer. It's like an intensive care unit. Nitrogen, phosphorus,
and potassium tend to be the three nutrients that get used up the fastest by rapidly growing crops and need to be supplemented throughout the growing season. And when the crops are harvested, most of
the vegetation is removed from the field, which also strips the soil
of nutrients. So agricultural crops—and possibly your vegetable garden—will use soil nutrients quickly and may require additional fertilizing through the growing season.

But most of our home landscapes are not dedicated to growing crops. Instead, we have permanent plantings: our lawns, trees,

shrubs, perennials, groundcovers, and bulbs. We're not growing them at a frantic rate because we're not harvesting them. When home landscape soils are tested, most of them have enough phosphorus and potassium—sometimes too much. With plenty of conventional and organic choices available to the home gardener, it's smarter to avoid the traditional NPK formulations and choose something that fits the needs of both your soil and your garden. Soil test reports often recommend specific fertilizers, or you can choose your own by carefully reading the labels and only buying what you need.

THE PROBLEMS WITH PHOSPHATE

Although excess levels of potassium don't seem to pose a problem—this element is fairly soluble and is quickly used by plants and microbes elsewhere—too much phosphate can wreak havoc on soil organisms, your plants, and, even worse, any nearby aquatic system. Let's look at problems right in your garden.

Excess soil phosphate, whether from organic sources like compost, bone meal, and bat guano or an inorganic source like rock phosphate, creates one of the most common nutrient problems in landscapes. You'll recall that iron deficiency in plants can be seen as interveinal chlorosis. It turns out that phosphate really does a number on iron, both in the plant and in the soil. First, it reduces the ability of plants to take up iron: the more phosphate in the soil, the less iron is taken up by the roots. Next, phosphate unfortunately combines with iron to create insoluble iron phosphate. This compound can't be used by the plant, nor can it be easily broken down. In fact, most soils contain plenty of iron, but it can't be taken up by the roots and it becomes unusable when there's too much phosphate.

If this wasn't bad enough, phosphate also inhibits the development of the mycorrhizal relationships between fungi and plant roots. These beneficial fungi are not able to penetrate the roots unless phosphate levels are low. So the gardener's best friends, the miraculous mycorrhizae, are MIA whenever phosphate levels are higher than necessary. The result? Your plants will need to expend more energy for root

Interveinal chlorosis in this rhododendron was caused by high soil phosphate levels.

growth than they normally would. In a sense, then, phosphate is stimulating root growth, but in a very bad way.

Ignore the seductive packaging that suggests supplemental phosphate will help you take home the blue ribbon for the biggest rose. As a rule of thumb, you should never add phosphate fertilizer to bulbs, groundcovers, perennials, shrubs, or trees unless a soil test states that phosphate is pretty much nonexistent. In your vegetable gardens, you may need to use it, especially if you are really pushing production. But follow the recommendations of your soil testing lab. Don't assume anything.

There! I've now saved you a chunk of change, as you'll never need to buy bat guano, rock phosphate, or bone meal again.

GREEN-WASHED FERTILIZERS

In those instances when you do need to provide specific nutrients to your plants, you'll have choices of inorganic or organic forms. Organic formulations aren't necessarily better, and many of these organic products, including bat guano, seaweed extracts, peat, and other exotic materials, deplete the natural resources of another ecosystem. For instance, the kelps routinely harvested for fertilizer use are literally the forests of the ocean. It's like clear-cutting an old growth forest simply to make an unnecessary product. Peatlands are the world's largest terrestrial repository of carbon, and degraded peatlands take centuries to regenerate after they are harvested. There's really no defensible reason for using these materials when good-quality substitutes are available. Environmentally conscious gardeners should avoid these green-washed products.

Most of the nutrients that you might need to add to your garden or landscape can easily and naturally be provided through organic mulches. Composts, wood chips, pine needles, and other recycled materials are not only good uses of materials that might otherwise end up in the landfill, but they also provide a slow feed of nutrients to the soil. This is the way that nature provides nutrients, and gardeners would be wise to follow this approach.

DELIVERING THE GOODS

There are always people with products to sell who recommend alternative methods of doing things. Fertilizer injection, either into the soil or directly into the trunk of a tree, is one of these methods. The idea is that by delivering nutrients directly to plant tissues, deficiencies are corrected faster and less fertilizer is wasted than by soil application. Science just doesn't bear this belief out. In fact, there are some serious problems with both methods.

Soil injection deliberately places fertilizer below the layers of fine roots found near the soil surface. Scientific testing has shown that this method is no more effective than soil application. It's a waste of money. Jabbing needles into trees and shooting them up with fertilizer is not only useless, it breaches the tree's protective barriers and creates nice pathways for insects and pathogens to enter the tree.

Another odd practice that seems to be gaining popularity is foliar fertilization, or spraying leaves with liquid fertilizer. There is some logic behind this notion, because plants can take up some nutrients through their stomata. Most commonly, fossil fuel gases containing sulfur and nitrogen can be taken in along with carbon dioxide and used in leaf proteins. The drawback to foliar fertilizers, however, is that some nutrients such as iron are immobile once they've entered a cell. These nutrients won't be passed on to other parts of the plant. And other nutrients, such as nitrogen, are required in such large quantities that foliar absorption can't possibly meet the entire plant's needs.

Foliar treatment is useful for diagnosing nutrient deficiencies, because you can test a small part of the plant and confirm whether there's something missing. Misting a chlorotic rhododendron leaf with iron chelate, for example, allows iron to be taken up. If the chlorosis is due to iron deficiency, the leaf will green up in a matter of days. But that's the only practical benefit of foliar sprays.

Don't substitute foliar fertilization for natural root uptake of nutrients. In many cases, foliar nutrient deficiencies are the result of an imbalance of soil nutrients, as we saw with phosphate and iron interactions. So applying fertilizer to the leaves doesn't address the real issue, which is a soil fertility imbalance. It's just a quick fix that treats the symptom and masks the underlying problem. It's like putting a Band-Aid over a splinter in your finger. You can't see the splinter anymore, but it's still there!

TEA? NO THANKS!

I could hardly write a chapter on plant nutrition without addressing the increasingly popular use of compost tea. For those not familiar with the product, it's made by adding compost to water and aerating the mixture, sometimes with additives like molasses, packaged microbes, or humic acids. It's appealing because it makes us feel nurturing and turns plant nutrition into a recipe. Not surprisingly, compost tea is heavily marketed as a natural way to fertilize plants by soil or foliar application.

I've been reading the science on compost tea for well over ten years now, and there is nothing that would encourage me to recommend its use. Nutrient and microbe levels are quite low, many times lower than what you'd find in compost alone; the process simply dilutes the nutrients and microbes.

Compost tea recipes can best be described as artisan, with their esoteric ingredients that sound impressive but are nothing more than marketing hype. This is the ultimate green-washed product: it has the environmental bells and whistles, but it wastes resources (all those pointless artisan ingredients) and expends energy. The aeration must be continuous and it requires electricity. Enthusiasts must apply the brew often, because it doesn't stick around. I'm a cheap-and-easy kind of gardener, so I can't imagine using such a product without any beneficial results.

My advice? Use compost, and let nature make the tea. Water will percolate through layers of organic mulches, giving your soil and plant roots a slow feed of nutrients. Not only is this cheap and easy, it's also natural and based on scientific evidence.

Ideally, our garden plants are well watered and full of essential nutrients. Now what? How does a plant take these inert chemicals and create sugars, which are the ultimate energy source for all those complex biochemicals? It's time to let a little sunlight in on the discussion.

Transforming Sunlight into Sugar

4

IT'S MIDSUMMER. The cool, wet weather of spring is a distant memory. Insects whir and buzz, and the garden overflows with plump leaves and flowers. But it's a different story when we walk over to the lawn. To conserve water we try to limit irrigation, but the thirsty nature of grasses means that they look a little sad and defeated in the midday sun. The exception are those annoying lawn weeds, such as crabgrasses (*Digitaria* species), purslane (*Portulaca oleracea*), and nutsedges (*Cyperus* species). They seem to defy drought and heat, looking like little green islands in a sea of yellow and brown. How can these weeds tolerate the dog days of summer, while our desirable grasses literally go to seed?

The answer lies within a plant's biochemistry, specifically, its photosynthetic pathway. Gardeners know quite well that plants are solar powered: they use the sun's energy to drive a complex set of chemical reactions that transforms a common gas (carbon dioxide) into a

The green weeds in this brown lawn have a unique biochemical pathway that helps them survive hot, dry conditions.

When sunlight, with its full spectrum of colors, strikes a green leaf, all the colors are captured in the leaf pigments except for most of the green light, which is reflected.

solid form (carbohydrates). Every time you enjoy a salad from your garden-fresh greens or lick the juice from the season's first strawberries from your fingers, you are experiencing the delightful results of photosynthesis. It is, without a doubt, the most important biochemical reaction for life on Earth.

Harnessing the Sun

The first sunny day of spring is a tonic to plants and gardeners alike. Tomato seedlings in the greenhouse and potted bulbs overwintering in cold frames respond eagerly to the warmth as well as the biochemical kick-start that energy from sunlight supplies. In fact, sunlight consists of a wide range of energy including powerful, short-wavelength radiation (including X-rays) as well as weaker, long-wavelength radiation (such as radio waves). Within this broad spectrum is the narrow rainbow of visible light used to run many biological reactions, including photosynthesis. Sunlight that touches a leaf can be trapped (absorbed), be bounced back (reflected), or pass through untouched (transmitted). Only the absorbed light can be harnessed and put to work inside the plant. And for light to be absorbed by a leaf, it has to be captured by a pigment.

Two groups of pigments are used to trap light for photosynthesis: the green chlorophylls and the orange carotenoids. It surprises many people to discover that the color of a pigment is actually the result of the light that isn't absorbed, but instead is reflected and transmitted. So chlorophyll doesn't absorb green light well at all, but it is very good at nabbing the red and blue wavelengths. Likewise, carotenoids capture blue wavelengths but reflect the red and orange portions of the spectrum. The color you see is whatever color is not absorbed.

Let's look at the rainbow riot of colors in my own garden in this scientific context. The red spring tulips and summer daylilies absorb every color except red, which is reflected back to our eyes, and we see red. Likewise, the yellow daffodils, green hellebores, blue hydrangeas, and purple clematis absorb all colors except for the ones that we can see. And those pure white mock orange blossoms? They don't absorb any visible light, so we see all that white sunlight reflected back to us. A pure black flower would swallow every bit of visible light like a black hole, leaving nothing for us to see. (In reality, most black flowers have some tinge of red, blue, or purple.)

Creating Sugar out of Thin Air

Now back to our green leaves, busily trapping bits of sunlight (particles known as photons). Once a photon has been captured by a photosynthetic pigment, it's used as a biochemical booster in creating plant food. The most important step of this long and complicated process requires leaves to suck carbon dioxide from the atmosphere and attach it to a small, five-carbon organic compound. This creates a six-carbon compound that immediately splits into two three-carbon

Plants use the sun's energy to transform carbon dioxide and water into sugar and oxygen.

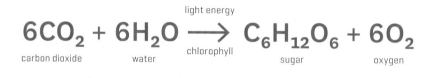

$$6CO_2 + 6H_2O \xrightarrow[\text{chlorophyll}]{\text{light energy}} C_6H_{12}O_6 + 6O_2$$

carbon dioxide   water   sugar   oxygen

compounds, which are biochemically poked and prodded until they end up in a sugar molecule. The term *carbon fixation* is used to describe this transformation of carbon from a gas to a solid form. All life on Earth depends on this reaction. (Well, except for anaerobic microbes whose dependence on other biochemical pathways has the unfortunate distinction of producing hydrogen sulfide and other Eau de Swamp gases.)

If photosynthesis is the most important biochemical process on Earth, then the enzyme that glues carbon dioxide onto that five-carbon compound is the most important enzyme. It has an import-ant-sounding name as well: ribulose 1,5-bisphosphate carboxylase oxygenase. Fortunately, plant physiologists love acronyms, and I'll call this enzyme RuBisCO. (Fun fact: RuBisCO makes up 30 to 50 percent of the total protein in leaves, making it the most common protein on a global scale. It's the letter E in the biochemical Scrabble game.)

RuBisCO is an ancient enzyme. The single-celled organisms that contained it evolved long before dinosaurs or cockroaches or even earthworms appeared on Earth. It's found in every photosynthetic organism, including bacteria and algae as well as higher plants. Earth was quite a different place when RuBisCO showed up, primar-ily because there was no free oxygen in the atmosphere. Life con-sisted of primitive microbes existing in oxygen-free (anaerobic) environments, like those that thrive in wet, poorly drained soils today. In much the same way, RuBisCO works best under anaerobic conditions, meaning it's able to do its job gluing carbon units effectively.

Oxygen: A Toxic Byproduct of Photosynthesis That We Can't Live Without

There's an important chemical reaction that occurs when leaves absorb photons for photosynthesis, and that's the splitting of water molecules. When water molecules are split, they release the electrons needed to drive photosynthesis and they also form molecular oxygen. Primitive Earth had no oxygen gas floating around, and this new chemical was a lethal, corrosive pollutant. As photosynthetic organisms evolved and oxygen began to accumulate in the atmosphere, a catastrophic extinction of early life forms followed. (Most modern anaerobes are now confined to underwater or underground habitats, where oxygen levels are low enough for them to survive.) New life forms appeared on Earth that could tolerate oxygen, and now most life on Earth is utterly dependent on a gas that was once a deadly pollutant.

Like most modern-day terrestrial organisms, plants not only adapted to the presence of oxygen but now use oxygen to break down carbohydrates and other organic compounds more efficiently. RuBisCO, however, continues to pose a problem for many plants because of its sensitivity to oxygen. Under certain conditions, this sensitivity can cause plants to die back—or just plain die.

Photorespiration: The Ultimate Weight Loss Program for Plants

Unfortunately, RuBisCO is rather sloppy in its ability to bind carbon dioxide for eventual attachment to that five-carbon molecule. When oxygen is around, RuBisCO preferentially binds it instead of carbon dioxide. (Kind of like choosing triple chocolate decadence over fruit on the dessert tray, even though the fruit's better for you.) Therefore, RuBisCO works most efficiently when leaf carbon dioxide levels are high and oxygen levels are low, temperatures are moderate, and water is plentiful. But botanical Fantasyland and reality don't always coincide.

When RuBisCO hooks up with oxygen instead of carbon dioxide, the plant's biochemistry shifts from photosynthesis to photorespiration, in which photosynthetic cells consume chemical energy and oxygen, producing carbon dioxide. Plants experience several negative outcomes when oxygen throws a monkey wrench into the photosynthetic machinery.

When oxygen is glued onto the five-carbon compound instead of carbon dioxide, a different five-carbon compound is created that can't be used in forming sugars. This deceptively simple chemical process causes several serious problems for the plant. First, RuBisCO is temporarily taken out of commission when it binds oxygen, meaning it can't bind carbon dioxide. Low levels of available RuBisCO mean slower photosynthesis. Second, the five-carbon compound that's usually bound to carbon dioxide has been wasted and has to be regenerated. Third, the regeneration process for making the five-carbon compound includes a step that releases carbon dioxide (the respiration part of photorespiration). Finally, all of these processes require energy to repair the problems caused by that single step of binding oxygen.

Unlike regular respiration, which all living organisms use as a way to create energy currency from stored food supplies, photorespiration uses up energy and generates useless chemicals that have to be reformulated. This is such a hugely negative process that plants in hot, bright, and/or dry environments can lose nearly half of their newly formed carbohydrates, which means they are losing weight. Think about struggling urban street trees in the middle of summer. Is it any wonder that many never seem to thrive?

Tweaking the System: C3 and C4 Plants

Then how do plants survive in environments that aren't cool, moist, and rich in carbon dioxide? Primitive plants were restricted to aquatic environments and forests, where photorespiration wouldn't pose much of a problem. In more hostile conditions, plants continually modify their biochemistry and sometimes their appearance, because they can't just get up and leave for greener pastures. Some plants crank out additional RuBisCO to swamp the system, whereas others tweak this enzyme to be more particular in choosing carbon dioxide over oxygen. By far the most interesting plants are those that dramatically overhaul basic photosynthesis and figuratively thumb their noses at photorespiration.

To keep things as simple as possible, I'll refer to basic photosynthesis as the C3 pathway because carbon dioxide is used to form two three-carbon compounds. All plants follow the regular C3 pathway, but some have developed extra biochemical steps to avoid photorespiration. The first of these steps binds carbon dioxide to a three-carbon compound, forming a four-carbon product, so this series of extra steps is called C4 photosynthesis. To fix carbon, the C4 pathway uses a completely different enzyme, phosphoenolpyruvate carboxylase, which I'll mercifully refer to as PEPcase. Unlike RuBisCO, PEPcase is strongly attracted to carbon dioxide and binds it tightly instead of oxygen, so the C4 pathway isn't affected by photorespiration. Unfortunately, the four-carbon compound can't be used in C3 photosynthesis, and eventually a molecule of carbon dioxide is clipped off deep inside the leaf.

Now if you've been keeping track of all these biochemical contortions, you might well wonder how these pre-C3 biochemical steps do a plant any good: it's just spent chemical energy fixing and then unfixing carbon dioxide. Where's the benefit?

A CARBON DIOXIDE MOLECULE SAVED IS A CARBON DIOXIDE MOLECULE EARNED

Let's look at some of the plants that have evolved C4 photosynthesis. They include more than half of all grass species as well as some herbaceous annuals and perennials, but no conifers. If you could crawl into a stomate on a giant C4 leaf and wander over to a vein, you would see it surrounded by a ring of tightly packed cells, known as the

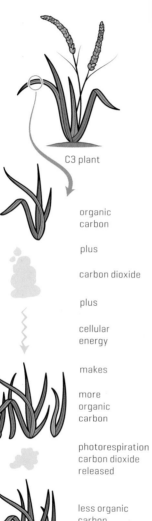

solar energy

C3 plant

C4 plant

organic
carbon

plus

carbon dioxide

plus

more cellular
energy

makes

more
organic
carbon

no photorespiration

same amount
of organic carbon

C3 plants use less cellular energy to fix carbon, but they lose carbon to photorespiration if conditions are not favorable. C4 plants use more cellular energy to fix carbon, but they don't lose it once it's fixed, meaning they continue to grow in conditions unfavorable for C3 plants.

bundle sheath cells. You'd have a tough time breathing here, because that tight packing of cells means very few air spaces and very little oxygen or any other gases for that matter. The low-oxygen environment bounded by the ring of cells is where RuBisCO is conveniently stored (remember, RuBisCO just can't resist oxygen). These rings are the final destination of the four-carbon armored cars, which release carbon dioxide once inside. So C4 plants have developed safe houses where RuBisCO can bind carbon dioxide without interference from oxygen, stymieing theft by photorespiration.

The biochemical brilliance of the C4 pathway is that carbon dioxide can be fixed into a solid form in hot, dry, and bright environments without interference from photorespiration. The four-carbon compounds are like armored cars shuttling precious carbon cargo to various parts of the leaf before releasing it again as carbon dioxide. The ultimate function of C4 photosynthesis is to accumulate carbon dioxide in oxygen-free tissues.

However, the cellular energy needed to run this carbon dioxide armored transport service is significant. It takes almost twice as much energy to make sugar by C4 photosynthesis as compared to the C3 pathway. In warm environments like tropical grasslands, the energy costs are minimal compared to the advantages of avoiding photorespiration. RuBisCO is protected from oxygen and binds carbon dioxide at maximum speed. C4 plants save energy by not manufacturing extra RuBisCO and by not having to remake the five-carbon compounds wasted in photorespiration. Finally, when the environment is hot, bright, and/or dry, the net gain of carbon dioxide fixed outweighs the energy costs of the additional C4 steps.

WHY C4 WEEDS ALWAYS WIN IN YOUR DRY SUMMER LAWN

Let's look at that dry summer lawn again. Bluegrass, fescue, and other cool-season C3 grasses that do so well in the spring (and in moderate climates in the winter) have gone dormant. The roots are still alive, but the blades have died back. Once day temperatures get much above

The interior leaf architecture of C4 plants allows them to concentrate carbon dioxide for photosynthesis.

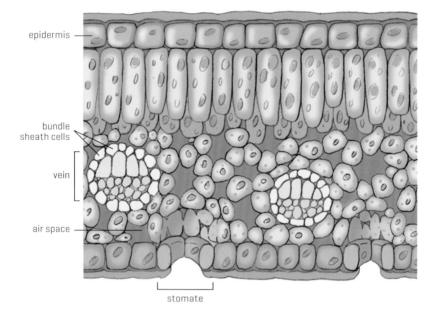

epidermis

bundle sheath cells

vein

air space

stomate

80°F these species take a break until cool moist conditions return. The C4 plants—the maddeningly healthy weeds we saw earlier—are in their element, photosynthesizing briskly with little interference from the dormant species. When temperatures drop and rains return, the extra photosynthetic steps of C4 plants become an energy liability instead of an asset, and the C3 plants return in full force.

As tenacious as crabgrass and other summer lawn weeds can be, even they can't survive extended summer drought. Though their photosynthetic mechanism is still functional, like most garden plants they are vulnerable to dehydration. Without some minimal water input, they too will go dormant.

Variation on a Photosynthetic Theme

My neighbor has a green roof and she never waters it, even in the summer. Peering over the gutter, I can see it's abundantly planted with different stonecrops (*Sedum* species) and hen-and-chicks (*Sempervivum* species). There's hardly any soil, definitely no shade, and yet it's a virtual ocean of succulent greenery. Surely these can't be C3 plants, and they don't look like any of the C4 weed species in the lawn. How do these little succulents survive and thrive in this environment?

A Baker's Dirty Dozen

Weeds that are C4 species can make a gardener's life miserable. Now you know why they do so well in the summer.

- Barnyard grass (*Echinochloa crus-galli*)
- Bermudagrass (*Cynodon* species)
- Black medic (*Medicago lupulina*)
- Cogon grass (*Imperata cylindrica*)
- Crabgrass (*Digitaria* species)
- Foxtail (*Setaria pumila*)
- Goosegrass (*Eleusine indica*)
- Johnsongrass (*Sorghum halepense*)
- Nimblewill (*Muhlenbergia schreberi*)
- Nutsedge (*Cyperus* species)
- Pigweed (*Amaranthus* species)
- Purslane (*Portulaca oleracea*)
- Spurge (*Euphorbia* species)

While the C4 photosynthetic pathway deftly sidesteps photorespiration, it gets hammered in perpetually arid environments. Leaf stomata must be open for carbon dioxide to enter, but then water vapor escapes. Shutting the stomata slows photosynthesis, and the C4 pathway becomes less efficient. Deserts, environments that receive less than 10 inches of rain per year, are characterized by succulent species in the cacti and euphorb families. First and foremost, desert plants must be able to survive drought. To do this they store water in fleshy tissues and reduce water loss by closing their stomata during the day. So, our little stonecrops collect carbon dioxide only at night when the stomata are open. But photosynthesis requires sunlight. How can biochemistry be reconciled with reality? Let's explore another fascinating side road off the photosynthetic pathway.

Gardeners and scientists alike have to marvel at how plants adapt to seemingly impossible conditions. Once again, the C4 pathway runs the carbon dioxide catch-and-release program, but this time it happens at night. (The PEPcase enzyme functions any time, day or night, and that works well for those green roof succulents.) Stomata open wide as the sun sets, allowing carbon dioxide to flow in as oxygen from photosynthesis flows out. The biochemistry in our succulents is exactly the same as for C4 plants: PEPcase binds carbon dioxide and creates the four-carbon compounds. But in succulent species these carbon dioxide carriers remain in the same cells as where they're produced, parked overnight in large, water-filled storage structures called vacuoles. At sunrise, the stomata close down, the carbon dioxide carriers pull out of their overnight parking lot, carbon dioxide is unloaded into the same cells where it was picked up a few hours earlier, and RuBisCO begins the C3 pathway in a moist, warm, carbon dioxide–rich environment.

Here's an interesting tidbit about the four-carbon compounds in C4 photosynthesis: they are all organic acids. This helps explain early observations on the Crassulaceae (meaning fat plant), the family that includes stonecrops and hen-and-chicks, which described them as having a "daily acid taste" in the early hours—kind of a cactus version of morning breath. It's the acid-storing property of succulents that

Well-watered euphorbs grown indoors often have leaves [LEFT], **whereas those in more natural environments do not** [RIGHT].

gives this photosynthetic detour its name: Crassulacean acid metabolism or CAM.

The one drawback to CAM photosynthesis is that the stomatal shutdown at sunrise leads to an internal oxygen buildup, so photorespiration increases as the day wears on. Therefore, CAM is inefficient and expensive compared to C3 and C4, and as a result CAM plants grow very slowly in their native habitats. Yet this pathway is widespread, found in thousands of plant species belonging to dozens of plant families including the agaves, cacti, euphorbs, lilies, and orchids. What unites all CAM plants is that their native environments all experience limited water or daytime carbon dioxide at some time during the year. CAM plants enjoy growth spurts during

Xeriscaping

THE PRACTICE
Using drought-tolerant species (xerophytes) for landscaping in arid climates.

THE SUPPOSED BENEFITS
Drought-tolerant species don't use much water, so homeowners will save money on irrigation.

HOW PLANTS RESPOND
Xerophytes are fascinating plants. They survive quite nicely with little water, but they are fierce competitors for water when it's available. In response to extra water (what we call luxury consumption), they reward gardeners with a flush of leaves or a burst of flowers. But once the water is gone, they often drop some or all of their leaves and go dormant. This is typical for plants in stressful environments. However, gardeners don't like this and will usually water the plants again to get that aesthetically pleasing response. You can end up using as much or more water keeping xerophytic plants lush-looking than it takes for drought-intolerant species. So by all means use xerophytic species, but realize you are creating a landscape that is naturally slow growing—unless, of course, you add lots of water.

Drought-tolerant plants are often water hogs.

the rainy season or when irrigated in landscapes, greenhouses, and indoors. That's why my elkhorn euphorb (*Euphorbia lactea* 'Cristata') only took fifteen years to reach 2 feet in our house, whereas it might take many decades to reach that height in nature. Abundant water plumps succulent tissues and can even encourage temporary leaves to pop out, only to die back when water becomes scarce.

Harnessing the Sun a Bit Too Well

Descending from my neighbor's green roof with a newfound appreciation for the tiny stonecrop, we spot a problem across the street. Last week the exceptionally neat homeowner in residence sheared his boxwoods into regulation shapes and some of the remaining leaves are unusually pale. Closer inspection reveals they're actually white. There are no signs of insects or disease. What could have happened to turn leaves white in the middle of summer?

Plants that love full sun can be pretty sloppy about collecting light. In fact, leaves exposed to full sun are generally small, thick, and have protective layers of wax or fuzzy hairs to prevent internal solar overload. My sun-loving garden herbs, lavender and rosemary, are good examples, with their thick needle-like leaves and waxy sheens. Or my sage plant, with its thick, woolly, lambs-ear foliage. In contrast, shade-loving plants like my giant hostas need to capture every stray photon to keep their photosynthetic machinery running. Their leaves are large and thin, with little surface protection. Species that tolerate both sun and shade develop leaves that adapt to whichever environment they are grown in. This last strategy mirrors the way that leaves develop in shrubs and trees: outer leaves are adapted to intense light conditions, whereas the shaded interior leaves are larger and thinner.

Interior canopy leaves are loaded with photon-capturing pigments to catch the scraps of light missed by the outer canopy. These interior leaves tend to be horizontally oriented and can be fairly long-lived. Shade leaves are important contributors to photosynthesis and should not be routinely pruned out as a means of tidying up a tree or shrub. To do so injures the plant's ability to nourish its interior canopy.

Speaking of pruning, we now have a good idea of why the neighbor's boxwoods have turned white. The interior leaves, once nestled in deep shade, were suddenly exposed to intense levels of full sun. The pigment-rich tissues absorbed much more sunlight than could possibly be funneled into the photosynthetic apparatus, and that extra energy has nowhere to go. It overloads the chloroplasts and literally fries them from the inside out, a process called photo-oxidative bleaching. Just as if they've been dipped in Clorox, the leaves are a pure white. All is not lost, however. Once the bleached leaves fall, new leaves will emerge that are adapted to full sun conditions. Remember this in your own garden, and save pruning for early spring or late autumn when sun injury is less likely.

Looking for That Place in the Sun

On the way back home, we automatically duck under the dogwood tree that leans out into the street. It started out straight enough when it was planted ten years ago, but has progressively grown away from the arborvitae hedge looming behind it. There are all kinds of tree bondage kits available for straightening up recalcitrant trees. Would one be worth a try?

True sun-loving species are garden divas: they are happiest in the middle of the solar spotlight. Surrounded by abundant sunlight, they capture only a fraction of the available radiation for photosynthesis. If moved off center stage into shaded conditions, they never reach their full potential and can languish for years waiting to be rediscovered by the sun.

It's easy to recognize these unhappy performers, especially when compared to their sun-drenched counterparts. Shaded sun-lovers have sparse leaves that are larger and paler than normal. The stem tissues between the leaves (the internodes) are longer as the plant puts extra energy into outgrowing the shade. Most obvious is the leaning growth habit that they will develop in an effort to reach the sun. The tilt is the result of an uneven growth pattern in which auxins cause the shadiest side of the plant to grow the fastest. The difference in growth rates causes the tree to bend away from the shade. Most of the leaves will be found on the least shaded side of the plant as well, emphasizing the asymmetrical shape.

But plant bondage kits won't help straighten unhappy sun-lovers. The plants will continue to reach for the sun. The only cure is prevention: put them on center stage far away from those who might steal the spotlight. Also consider using light-colored mulches and other reflective surfaces to add some extra brightness.

Sun-loving species try to escape the
constant shading of conifers.

Photosynthetic Parasites

Most of the plants we cherish in our homes and gardens follow the
typical photosynthetic lifestyle. But some do not. Thumb through any
wildflower guide and you'll find a small section on plants that aren't
green but instead sport white, brown, or yellow hues. These species
are parasitic because they siphon away sugars from hard-working
photosynthesizers without giving anything in return. If you were able
to burrow underground and follow their roots, you'd find these spong-
ing off the roots of green plants, usually with the assistance of some
fungal partners.

LEFT: **Pine drops are the flowering portion of an underground parasitic plant.** BELOW: **Indian pipe emerges through the forest duff layer.**

Even though their botanical habits may be unsavory, these parasitic plants are both unusual and beautiful. Indian pipe (*Monotropa uniflora*) and pine drops (*Pterospora andromedea*) are just two of these lovely species. Unlike their photosynthetic cousins, these plants are pure white and chocolate brown, respectively. And woe to anyone who illegally digs these plants to use in their own garden, for without their photosynthetic hosts and fungal go-betweens, these species will never survive.

Photosynthetic variants are unusual. We understand the importance of being green. But sometimes normally green leaves transform to the chromatic opposite of green: red. What's the trigger for this switch?

Why Leaves Can Turn Red Anytime, Anyplace

5   **I'M A LUCKY GARDENER,** because I live in a region with distinct seasons. I actually enjoy storing summer clothes and rediscovering what's in my winter wardrobe. Likewise, my garden packs away its languid summer greens, replacing them with bold reds, oranges, and yellows. If you're the fanciful type, you may like the notion of Jack Frost painting the autumn leaves. And indeed, cooler night temperatures speed up autumn color transformation. But wait a minute! Some plants have leaves that turn red in midsummer, some whose leaves are red when they're young and then turn green, and some whose leaves are always red. Why do leaves turn red, anyway?

Anthocyanins: Yet Another Fine Plant Product

It seems as though plants possess more pigments than an artist's palette! That's because light is one of the most important and reliable environmental signals plants receive. Pigments select and collect particular wavelengths of light for photosynthesis and other botanical business. The pigments that reflect and transmit most of the red, blue, and purple colors we see in leaves, flowers, and fruits are anthocyanins, a name that literally means flower pigment. Unlike chlorophyll and carotenoids, nested cozily in the chloroplasts, anthocyanins are water-soluble mobile pigments that can scoot easily within and between cells and tissues.

Anthocyanins are multipurpose molecules. Their mobility in water means they can be used as cellular transport systems. Sugars, metals, and other substances can be hooked on to anthocyanins and zipped off to other parts of the plant that need them. Anthocyanins are also a natural sunscreen: under full sun conditions they can protect chloroplasts from solar overload by siphoning off some of the incoming energy. They're also powerful antioxidants, meaning they can protect

The green-leaved rootstock of this laceleaf maple could take over the red-leaved scion if new green sprouts aren't pruned.

plant cells from all kinds of environmental stressors. This is why nutritionists recommend that we eat so many red and blue fruits and vegetables: these antioxidants are beneficial to us, too.

Why Aren't All Leaves Red?

With the advantages that anthocyanins offer plants, the curious gardener might wonder why all leaves aren't red. To answer this, let's look at some tree cultivars that have been deliberately selected for their permanent red foliage. There are several flowering crabapples with various hues of red foliage available at most nurseries, and many of these have been grafted onto a green-leaved rootstock. If we were to study these trees in the landscape, we'd come across several with green-leaved suckers threatening to overtake the red-leaved scion. Clearly, the rootstock of these crabapples is more vigorous.

It turns out that anthocyanins are expensive to manufacture. Their synthesis takes chemical energy that the plant could otherwise funnel into growth. Even worse, these pigments can interfere with photosynthesis, further reducing the plant's energy reserves.

Okay, now it sounds like I've just contradicted myself. How can anthocyanins interfere with photosynthesis if they're not in the chloroplast? Let's review what pigments do with light. Chlorophyll looks green because it absorbs red and blue wavelengths, reflecting and transmitting the green wavelengths to our eyes. The absorbed light is what's used for photosynthesis. Red anthocyanins reflect and transmit red light, while they absorb blue and green light. The absorbed blue light doesn't do any work in the anthocyanin molecule—it's not making sugars or something else—but it's not available to the chloroplasts, either. So red leaves aren't able to use all of the incoming blue light for photosynthesis, and the leaves' ability to make sugars is stifled. Green leaves naturally have the advantage over red leaves in cranking out carbohydrates. In fact, research on red- and green-leaved *Coleus* has shown exactly that. And red-leaved cultivars grown in the shade tend to be more green than red.

The Young and the Stressless

Tender, emerging leaves are at the mercy of an often hostile world. Insects wait to feast on them. Sunlight can parboil them. Diseases riddle them until they resemble lacework. One of the ways some plants protect their new leaves is with a healthy dose of anthocyanins. This temporary condition, called juvenile reddening, is usually most intense in leaves exposed to full sunlight. Part of this is because of that natural sunscreen function I mentioned. But there may be another benefit.

Leaves are pretty darn tough when they're mature. But when they are young and expanding, they don't have the tough waxy cuticles,

Watering and Leaf Wilt

THE PRACTICE
Interpreting leaf wilt as a signal to add water.

THE SUPPOSED BENEFITS
Leaf wilt means that water levels are low in the leaf, so that must mean they are low in the soil, too.

HOW PLANTS RESPOND
While it's true that leaf wilt is generally due to lack of water in the leaf, the underlying cause is not necessarily lack of soil water. What leaf wilt does tell you is that the roots are not doing their job in taking up sufficient water. Soil that's too wet, too dry, too hot, too cold, compacted, or salty can injure or kill roots. Pathogens, insects, and rodents can do the same. Poor root systems—those that circle, kink, and generally tie themselves in knots—don't take water up well. You, the gardener, realize there's a problem when the leaves droop like steamed spinach. Root around the roots a little and try to diagnose the trouble. Don't just turn on the hose!

Leaf wilt can be caused by many factors, not just lack of water.

Browning on the tips and
edges of leaves is an indicator
of drought stress.

rigid cell walls, and the other natural defenses that older leaves do. Not only does this make the leaves easier to attack from the outside, but it also means that water loss is greater, too. We can see this happening in the garden on a hot summer day. The newest stems and leaves on our plants wilt first, because they can't control the evaporation through their expanding surfaces. If water uptake from the roots doesn't keep up with demand, then these young leaves lose so much water that their tissues are killed. Usually this happens first at the tips and along the margins of the leaves, where you'll eventually see browning (or necrosis) as an indicator.

Now consider some common garden plants that show juvenile reddening, such as *Photinia* and English ivy (*Hedera helix*). The young,

If water is limiting, new leaves won't expand to their full size.

expanding leaves rarely wilt, and I've never seen leaf browning from water loss in either plant. In this case, what anthocyanins might be doing is holding water inside the leaves by adhesion. Remember, the general rule is "Water always moves to where water isn't." Because red leaves have anthocyanins swimming around in the cellular water, that means they have relatively less water than green leaves. It's harder for water to evaporate away from red tissues than from green ones.

A final problem young leaves face is successfully reaching full size. During this brief window of time, the bonds between cell wall fibers are loosened, allowing them to slip past one another. When leaves are expanding, they rely on the power of water—turgor pressure—to force cell wall expansion and reach mature size. If young leaves are not turgid enough, whether it's from too little water uptake or too much water loss, they won't reach maximum size. In fact, the presence of notably smaller leaves on evergreen trees and shrubs is a good indicator of drought stress.

Not only can young red leaves retain their water in cellular lockdown, they also attract more water from adjacent tissues. Increasing the water content plumps up young leaves, creating the turgor pressure needed for cell walls to slide past one another and guaranteeing maximum leaf size. Once the leaves have stopped expanding, the anthocyanins disappear and red fades to green.

How Dry I Am, How Red I'll Be

While some plants have red leaves only during their teenage years, others wear these flashy pigments all their lives. Many of these are species found in environments that are bone dry much of the year.

We've already seen that anthocyanins help young leaves hang onto their water during expansion. In the same way, red-leaved trees and shrubs can tolerate droughty conditions any time during the growing season. When rainfall peters out and soils hoard their water, many plants have a tough time just surviving, much less thriving. But anthocyanins can help in the same way they help juvenile leaves.

Usually drought-tolerant plants have small, thick leaves to resist

Water Crystals (Hydrogels, Dry Water)

THE PRODUCT
Water-absorbing chemicals are added to soils during transplanting or when normal irrigation isn't possible.

THE SUPPOSED BENEFITS
Water crystals absorb large amounts of water and slowly release it into the surrounding soil as it dries.

HOW PLANTS RESPOND
Hydrogels are very good at retaining water, and they do release it slowly as plant roots take up water from the soil. But there will come a point when the gel won't release any more water, and instead will suck it up from the surrounding soil, as well as from plant roots. So using hydrogels is not a good substitute for normal watering. Using them around new transplants, with their actively growing roots, is a bad idea. In fact, research has demonstrated that these gels aren't any more effective than just using a decent organic mulch to help keep the soil hydrated.

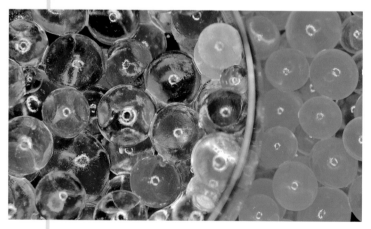

Hydrogels not only provide water to roots, they can suck it away as well.

water loss. Red leaves, however, can afford to grow larger because the anthocyanins help collect and maintain a healthy water reserve. Here's where plant physiology and ornamental value happily coincide. Research on red-leaved cultivars of *Viburnum* and smokebush (*Cotinus*), for instance, has shown that these attractive choices require less watering than green-leaved cultivars. The same is probably true for many other red-leaved species whose drought tolerance has not been scientifically studied. Be prudent is using these red-leaved cultivars, however. They're not good choices for well-watered sites, where green-leaved cultivars are more competitive.

Droughty soils can stress plants beyond just denying them water: they often contain high levels of salts. The more evaporation that occurs, the greater the concentrations of salts become. Whether they occur naturally or as a result of excessive fertilization, these salts can doom sensitive plants.

A Salty Tale

As a youngster growing up in the sometimes soggy Pacific Northwest, I was often sent outside with the salt shaker to kill slugs in my mother's garden. (Gives a whole new meaning to "salt of the Earth," doesn't it?) It was with a mixture of horror and amazement that I would watch the slug shrivel before my eyes. Later I would find a miniature carcass left where a plump slug had happily munched petunias only hours before. Though I no longer use the seasoning method of slug control, the experience taught me an unforgettable lesson about how salt affects water movement.

Once more, we just need to remember the water movement rule: "Water always moves to where water isn't." Salt dissolves in water, so adding it to the slimy skin of a slug draws water out of the slug's body. In much the same way, salt in the soil can pull water away from plant roots, making it more difficult for the plant's internal irrigation system to function.

Once again, it's anthocyanins to the rescue! Because anthocyanins help plant tissues hang onto their internal water, a tug-of-war

The tiny glistening dots on this ice plant [TOP] are salt storage bladders [BELOW].

between salt molecules and anthocyanins ensues. We can find anthocyanins in the roots, stems, and leaves of plants adapted to salty environments. Remember, it's not just the roots that need to pull water out of the soil. The leaves need to pull water from the roots as well.

Halophytes (literally salt plants) are exquisitely adapted not only to tolerate salty conditions but to revel in them. The species scientists have studied so far appear to use anthocyanins to hoard internal water. Mangrove trees, which inhabit salt marshes, collect anthocyanins in their leaves. In desert environments, ice plants such as *Mesembryanthemum crystallinum* collect anthocyanin-like pigments in bladders at the tips of their leaves. These little red bladders pull water from the soil, through the roots, and into the leaves, leaving the salt behind. It's likely that other plants living in salty soils or in coastal regions have the same ability to use anthocyanins in extracting and preserving water.

A Ripe (and Red) Old Age

We've seen some definite advantages for young leaves and mature leaves to hoard anthocyanins, but what about old leaves? Specifically, what about all the trees and shrubs whose leaves turn red in autumn before they finally die and fall off? They're certainly not expanding any more, and the chance of drought stress is remote. Yet every autumn both gardeners and non-gardeners alike wonder what causes leaf color change and why it varies so much between different species and different climates.

First, let's consider what deciduous woody plants are experiencing in autumn. The days are often bright and warm, and photosynthesis chugs right along. Evenings, however, are markedly cooler and on clear nights we start getting light frosts. The foliage of deciduous species, unfortunately, isn't able to withstand severe freezing, so trees literally cut their losses by dropping these leafy liabilities. It's a delicate balancing act trees must perform: on one hand they need to photosynthesize as far into autumn as possible, but on the other they must mobilize leaf nutrients and shuttle them elsewhere for winter storage

Sugar maple leaves change colors with the onset of cooler autumn temperatures.

before killing frosts move in and leaf nutrients are left stranded in dead foliage.

The most protected parts of trees and shrubs are the large woody tissues—branches, trunks, and roots—and this is where nutrients are tucked away until spring. It's fairly easy for leaves to break down proteins into amino acids and starch into sugars, and these smaller building blocks are easier to transport through the plant. But individual sugar molecules, such as glucose and fructose, are reactive and have a tendency to bind to other reactive compounds inside cells. To prevent molecular mayhem from gumming up the works, leaves use carrier molecules to bind the sugar and carry it away. What are these carrier molecules? Anthocyanins, of course!

A combination of short days, cool nights, and increased sugar levels flips the switch on the anthocyanin assembly line. Newly minted anthocyanins are paired up with sugars and begin their seasonal migration, just like the snowbirds heading for points south. At the same time, other salvageable leaf structures are being broken down and transported for storage: anything that can be recycled will be. As photosynthesis grinds to a halt and sugar levels decline, the levels of anthocyanins decrease, too. Eventually, all that's left of the leaf is a dried brown shell.

Why are anthocyanins so important in this process? In other words, why isn't there some other carrier molecule? It turns out that anthocyanins are providing double, maybe even triple, duty in autumn leaves. They transport sugars, they help the leaves retain water, and their antioxidant activity may even protect the leaves from environmental damage from daytime sun or evening frost during this sensitive stage.

You may know how important antioxidants are to human diets. They're also important to the plants that manufacture them. Anthocyanins are incredibly powerful in this regard; they're even stronger than vitamin E. As leaves begin to die off, the protective cuticles erode and the internal tissues become more sensitive to sunlight and other environmental factors. The stress that these factors cause results in the formation of reactive oxygen radicals and other potent, destructive molecules that poke holes in membranes and generally cause death and destruction. Anthocyanins neutralize these radicals, allowing the leaves to stay active until nutrient transportation stops.

The coolest thing about this entire shutdown process is that we can see it happening. Let's look at a sugar maple (*Acer saccharum*) and its stunning display of autumn color. First, the green leaves seem to turn bright red overnight, as the anthocyanin production machinery turns on full bore. While anthocyanins transport sugars, the pigments in the chloroplast are being disassembled for parts. The leaves lose their greenness, and we see the bright red-orange combination of anthocyanins and carotenoids appearing at the margins and working its way inward. The carotenoids disappear next, and the leaves turn yellow as the last chemical stragglers are swept out of the leaf and into their winter destination. The connective tissue between the leaf and the stem is sealed off, and the leaf dies. The stage is now set for winter's arrival.

It's a Cold, Cruel World

Winter in the garden can be both beautiful and desolate. Songbirds have migrated to escape potentially deadly freezes. Squirrels and other winter visitors brave the cold with thickened fur, layers of fat, and other protective methods. Plants have a variety of winter survival tactics, too. Annual flowers and weeds simply escape winter altogether, flowering and dying before temperatures turn cold and leaving their seeds behind to carry on the following spring. Hostas and other herbaceous perennials lack leaf protection against cold and die back to the root crown insulated in soil and mulch. The buds and woody tissues of trees and shrubs have chemical and physical barriers to ice invasion, but leaves do not. As we've seen, deciduous species shed their leaves, but what about evergreens like rhododendrons, cedars, and ivy? Not only are the leaves of these species still alive, they keep churning out sugars all winter long. They're full of water, so why don't they freeze solid like birdbaths and garden hoses?

Well, let's look at those birdbaths and garden hoses first. Their water is relatively undiluted, meaning the water doesn't have a lot of dissolved sugars or other chemicals in it. When ice first starts forming, it begins a chain reaction: wherever ice touches liquid water, that water is transformed into ice. Ice formation is slow at warmer temperatures, say just around freezing, but as the temperature drops, freezing is faster. (The freezing rate is important when it comes to predicting plant damage, and we'll come back to this topic in a bit.)

First we're going to do an experiment, one that is easy enough for you to do at home. We'll start with two very clean, narrow glass containers like test tubes. After filling them about halfway with water, we brush a little dust into one of the tubes and put them both in a freezer. After an hour or so, we check on them to see if they've begun to freeze. The one we sprinkled with dust is starting to get an icy film on the surface, but the other is still in liquid form. We check on the first one in another hour and can see ice has grown along the inside surface of the tube, but there's still a core of liquid water left in the middle. The other test tube is still unfrozen. Finally, after three hours the first tube has filled with ice, but the second is still liquid. The temperature of the water in this second tube is actually below freezing, maybe around 28°F. This is known as supercooled water.

Now the fun begins. We take a toothpick and touch the surface of the water in the second tube. Immediately the entire tube freezes solid, so it looks just like the first tube. The supercooled water was waiting for a nucleator—in this case the toothpick—to start the chain reaction. We've just observed the differences between slow and fast freezing events

Before we move on, let's consider why we only filled those tubes halfway up. You know what would have happened if they were brimming with water: the ice would have expanded over the edge and made a mess. (Water is unusual in that its solid form has lower density than its liquid form and therefore takes up more space.) If you're as absent-minded as I can be, you too may have put a can of pop or bottle of wine in the freezer to cool it quickly and then forgotten about it until much later. You know what you're going to find when you finally open the freezer again!

Now, let's substitute plants for the test tubes and see what happens. When water freezes quickly inside of plant cells, the expanding ice stretches the cell membranes until they split open. When you place lettuce or other greens too close to the cooling elements in a refrigerator you'll see the results: soggy, glassy-looking leaves. The same can happen in your garden, though it's rare for temperatures to decrease that rapidly. Instead, what usually happens is that ice forms slowly in the leaves, particularly in the air spaces outside of the cell walls. This intercellular space contains water and is surrounded by nonliving tissue, so when the water freezes it doesn't cause any damage. So far, so good.

Remember, "Water always moves to where water isn't." As far as

Plant cells contain dissolved substances that keep cellular water from freezing, encouraging ice to form between the cells instead.

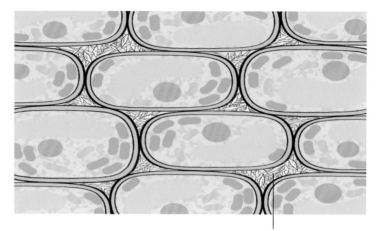

ice crystals

the plant is concerned, ice isn't water. Ice can't be taken up by roots or dissolve nutrients. The intercellular spaces become drier as ice forms, sucking water away from the nearby cells. Now the living cells in the leaf are dehydrating and becoming water stressed. The longer the freezing in the air spaces continues, the more dehydrated the living cells become. If it goes on too long, the cells eventually die. But they die from lack of water, not from ice damage.

How do evergreen leaves combat this frozen doom? You know the answer by now: it's anthocyanins! Low temperatures cause leaves to accumulate anthocyanins in a wide variety of garden favorites, both evergreen and deciduous, including maples (*Acer* species), red-twig dogwood (*Cornus sericea*), smokebush (*Cotinus*), *Euonymus* species, English ivy (*Hedera helix*), Oregon grape (*Mahonia repens*), apple (*Malus domestica*), pines (*Pinus* species), peach (*Prunus persica*), and sumac (*Rhus* species). We've already seen how anthocyanins help deciduous leaves prepare for their autumnal demise, but evergreen leaves accumulate anthocyanins as a kind of antifreeze. Once accumulated, anthocyanins help red-tinged leaves of Oregon grape, viburnum, and rhododendron hang onto their cellular water better than green leaves of the same plant. Water that's bound inside the cells is not able to migrate easily into intercellular spaces, so the leaves are able to withstand colder temperatures. They can accumulate so many anthocyanins that their leaves are visibly redder than they were during the summer. When spring returns along with warmer temperatures, winter-reddened leaves regain their usual green color. This doesn't happen in all evergreen species, but those that do turn red sure brighten up our winter landscapes.

This freeze-avoiding benefit that anthocyanins offer is also useful in those juvenile leaves we talked about earlier. Leaf buds are tough little structures, resistant to hard winter freezes, but once they break dormancy the emerging leaves are susceptible to late spring frosts— even mild ones. Ice forms quickly on top of and inside these tissues, and that's the end of those leaves. But anthocyanins can lower the freezing point of water, just like adding salt to sidewalks keeps them from freezing. Juvenile anthocyanins allow leaves to survive their youth, much like immunizations help our children avoid fatal diseases and become teenagers. (And we have the pleasures of gardening to help *us* survive teenagers.)

Anthocyanins and Heavy Metals

I'm continually amazed at how anthocyanins can protect plants from so many different environmental insults. And few environmental insults can have as severe consequences as those of heavy metal contamination.

Heavy metals are a loose collection of mostly metallic elements with significant effects on life. Some heavy metals, such as iron, are nutritional requirements of animals and plants alike: we can't live without them (at least in small quantities). Others have no known benefit and in large enough quantities can be toxic. Many gardeners are well aware

of this problem, testing their soils for lead, arsenic, and other common heavy metals before planting vegetables and other edibles.

One of the dangerous things about heavy metals is that, because they are elements, they don't break down to nontoxic forms. Worse, they can be accumulated over time. We're told, for instance, to limit our consumption of swordfish because it can be high in mercury. So, the more swordfish we eat, the more mercury we accumulate. In high enough doses, mercury and other toxic heavy metals start to interfere with our enzyme activity and eventually can be fatal. Therefore, we try to limit our exposure to these metals as best we can, such as getting rid of lead in gasoline and paint and removing arsenic from wood preservatives and pesticides.

Because plants are stuck wherever it is they're growing, they adapted to heavy metals long ago. For some species, this means not growing where toxic heavy metals exist. But other plants tolerate and sometimes concentrate these elements, and because they have few competitors they grow profusely. There are arsenic-accumulating ferns, mercury-mining mustards, and cadmium-collecting crops. They employ different survival strategies, sometimes avoiding metal uptake altogether and sometimes taking it up but isolating it somewhere far away from sensitive enzymes. We can think of this last strategy as putting toxic heavy metals in time out.

Part of keeping heavy metals locked away is to make them less toxic to the cells, and this is where anthocyanins come in. Anthocyanins bind to reactive compounds such as sugar, and heavy metals are also reactive. Researchers have discovered plant species that naturally take up and bind heavy metals to anthocyanins, detoxifying the metal and allowing the plants to survive on these hostile soils. The flowers of one such plant, *Linanthus parviflorus* (a member of the pink family), turn from white to pink when exposed to soils containing heavy metals! It's like a botanical soil testing system.

A Watery Conundrum

We're going to step into my front yard and look at a Korean dogwood (*Cornus kousa*) that I planted eight years ago. This lovely tree was my poster child touting the benefits of wood chip mulches. But an odd thing happened after the first five years: the leaves became progressively smaller, redder, and less numerous than previous years. Finally, we realized that the tree was going to die unless we took drastic steps. When we dug it up, we found a perched water table (thanks to the house builders who buried clay excavated from the foundation 30 years ago) and a root system that had rotted away to nearly nothing. Why does waterlogged soil turn leaves red?

The pore spaces in garden soil, the areas between soil particles, are filled with either liquid (usually water) or gases (usually carbon dioxide and oxygen). So any time there's standing water in your garden soil, you can bet there's little oxygen. This oxygen deficit, or hypoxia, can doom plants that aren't adapted to wet soils, like my Korean dogwood.

What causes hypoxia? Sometimes it's flooding or a perched water

Aeration Tubes

THE PRODUCT
Plastic pipes installed vertically into tree planting holes.

THE SUPPOSED BENEFITS
Pipes bring oxygen down to roots like a snorkel.

HOW PLANTS RESPOND
In unobstructed landscapes, tree root establishment is primarily horizontal, as surface soils are most likely to contain substantial oxygen, water, and nutrients. In an enclosed area such as a container, roots contact the wall and grow deeper into the soil. An aeration tube may work in such a small volume of soil. However, a single pipe is unlikely to provide enough oxygen in the landscape, and field research has not supported the use of aeration tubes in outdoor settings.

Root snorkels aren't effective in aerating soil or enhancing tree survival.

Contact between different soil types restricts water movement, leading to a perched water table.

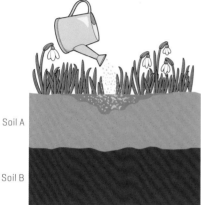

Soil A

Soil B

time

$\longrightarrow$

table, created when soils of different types are layered onto each other. Either situation will fill those pore spaces with water. Compaction from vehicles, foot traffic, heavy equipment, and even pets can squeeze the oxygen (and the life) out of soil. Heavy clay soils, with their fine particulate texture, have correspondingly tiny pores and little space for oxygen.

Now let's look at my dogwood's pathetic root system. All but one of the woody roots had rotted away, and the remaining system consisted of fine roots near the soil surface where oxygen could still be found. This shrunken root system could hardly support the substantial crown that developed in the first five years. But unlike the situation in most drought-stressed trees, the leaves on my waterlogged tree didn't die or even become brown. They were just small and red.

Our reliable, multipurpose anthocyanins were on the job again. Even though water uptake by the roots was a fraction of what it should

The bulls' eye pattern on these salal
leaves is due to fungal infection.

have been, anthocyanins prevented evaporation from the leaves. Compared to soils with plenty of available oxygen, researchers have documented increased anthocyanin production in the leaves of trees in hypoxic soils, including apple (*Malus* species), pear (*Pyrus* species), madrone (*Arbutus menziesii*), and red maple (*Acer rubrum*). Generally, once the problem is solved (the floods recede or you transplant your unhappy plants), the leaves turn green again as the anthocyanins are broken down. New leaves are bigger and healthier with improved root water uptake. And that's exactly what happened to our Korean dogwood when we moved it to better-drained soil.

Pushing the Red Button

Science has only grazed the tip of the anthocyanin iceberg. It's likely that these leaf pigments are instrumental in protecting plants against a slew of other environmental insults. For instance, leaves may turn red when they're short of certain nutrients, such as boron, magnesium, nitrogen, phosphorus, sulfur, and zinc. It's unclear what anthocyanins are doing in these situations, but because so many nutrients are associated with controlling water levels in the plant, anthocyanins may reduce water loss when these nutrients are deficient.

When soil pH is lowered beyond what's normal for plants in that area, red leaves may develop in response. Because acidic pH makes heavy metals more available for root uptake, it's likely the anthocyanins are binding those metals and storing them away from sensitive enzymes.

Finally, what gardener isn't intimately familiar with the bright red bulls' eye pattern caused by fungal disease on leaves? The center is the original point of infection, which dies and turns brown as the fungus invades the leaf. The reddest part of the pattern is on the battlefront between healthy leaf and fungal invaders. Are the anthocyanins actively fighting the fungus, perhaps using their antioxidant powers? Or are they just helping reduce water loss as the leaf tissues are exposed to the elements? Because we may also see this reddening in response to bacterial attack or other wounding, it may be an attempt to reduce dehydration along those damaged edges.

I guess I've exposed my plant biochemistry geekiness in spending an entire chapter showcasing anthocyanins. We know that they can tell us about soil conditions, water availability, and even the changing of seasons. And I firmly believe that science will continue to demonstrate that these pigments are major players in many plant defensive moves. Savvy gardeners can use this easy visual cue to figure out what environmental factor plants might be responding to, and what—if anything—they should do about it.

As you might guess by now, I think anthocyanins are the coolest thing in the plant pigment world. They can tell us about soil conditions, water availability, and even the changing of seasons. But autumn reds are a response to winter's approach. They aren't part of a plant's internal clock. So how do our deciduous trees know when autumn is in the air?

How Plants
Tell Time

6

LIKE CLOCKWORK, the butter-and-cream daffodils in my garden are the first flowers to open, followed quickly by a painter's palette of tulips, crocus, and irises. As spring expands into summer, lilacs and rhododendrons give way to gladiolas and asters. Flowers like clematis make a quick but stunning appearance, whereas others seem to flower all summer long, like my heavenly scented daphnes and gardenias. Taking the lead from the U.S. Postal Service, neither snow nor rain nor heat nor gloom of night stays these couriers from the swift completion of their appointed rounds. How do plants know when to grow and flower? How are they able to tell time?

Paul Simon begged mama not to take his Kodachrome away, and plants could make the same plaintive request about their phytochrome. Just like film, phytochrome changes when it's exposed to light, and it's these biochemical changes that allow plants to tell time. It's because of phytochrome that plants know when to burst into bloom, when to fan out those new leaves, and when to pack it up for the year and go dormant.

Yes, we're going to get into biochemistry again. But this time you don't have to worry about complicated reactions and enzymes. We're going to discover how different colors of light can transform phytochrome, and we'll follow the amazing changes it causes in plant behavior.

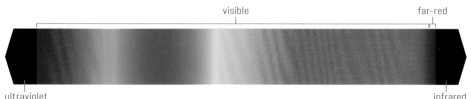

Sunlight includes ultraviolet, visible, far-red, and infrared radiation.

Phytochrome: A Pigment for All Seasons

Phytochrome literally means plant pigment. Obviously it's not the only one, but apparently people who name pigments ran out of clever ideas after naming chlorophyll (meaning green leaf), carotenoids (meaning carrot-like, referring to the orange color), and anthocyanins (meaning blue flower). This pigment is found in leaves but is not involved at all in photosynthesis. Instead, its value lies in being a shape-shifter and triggering some important changes inside the plant.

When phytochrome sits in the dark for a period of time it slowly reverts to a form that's called phytochrome red or P_r for short. The name refers to this form of the pigment's affinity for red light. After red light has been absorbed, the pigment quickly morphs into another form called phytochrome far-red or P_{fr}. As before, this is shorthand for what type of light this form of phytochrome absorbs. You can think of phytochrome as nature's little mood ring, but it's sunlight that triggers the response rather than emotions.

Okay, what the heck is far-red light? Let's visit my spring garden to figure this one out. The sun's shining, though scattered rainclouds threaten to drive us back inside. Suddenly a rainbow appears, the happy result of sunlight being segmented into distinct bands of color by prismatic raindrops. We're able to see the range of color from red on top to violet below, but there are also wavelengths beyond what our eyes perceive. On the inner edge of the violet arc are the invisible ultraviolet wavelengths, while the outer red arc of the rainbow fades into the aptly named far-red wavelengths. We can't see them, but plants can sense their presence.

Far-red light is not used for photosynthesis, although red light definitely is. And it's exactly this fact that makes far-red light such a useful indicator for plants. Leaves contain lots of phytochrome molecules, and what's important is the ratio of its two forms. If there is abundant red light—in other words, if full sun is striking the leaf—then most of the phytochrome will be in the P_{fr} form because red light converts the P_r form to P_{fr}. On the other hand, if there's more far-red light than red light available—for instance, red light has been captured by a canopy of actively photosynthesizing leaves above the

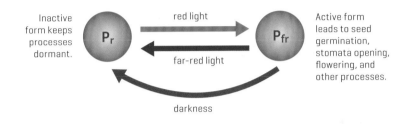

Phytochrome changes form depending on what type of light it absorbs. When the plant sits in darkness for some time, P_{fr} reverts to P_r.

Inactive form keeps processes dormant.

red light

P_r

far-red light

P_{fr}

Active form leads to seed germination, stomata opening, flowering, and other processes.

darkness

plant of interest—then phytochrome will be primarily in the P_r form. Put very simply, phytochrome tells plants whether there is enough of the right kind of light needed for photosynthesis, and plants respond accordingly. Let's look at some of these responses in the garden, starting with every gardener's bane: weeds.

Photodormancy

Though I put a thick layer of nice, chunky organic mulch down every year in my landscape, there's always a patch or two where the wood chips get dislodged. I don't notice until a crop of annoying weeds seems to explode from this bare patch of soil. I know the weed seeds were there all along, so why didn't they grow through the mulch?

Many weeds, as well as desirable plants, have fairly small seeds. This is an important trait for immobile organisms like plants, which don't have an easy way to farm the kids out. Small seeds are more easily blown by the wind or carried by water away from the parent plants. But small seeds also have small food reserves, and woe to the seedling that starts growing without enough sunlight to get the photosynthetic machine cranked up before reserves run out. These small seeds

When mulch layers become too thin, light will cause weed seeds to germinate.

tend to be photodormant, meaning they will not grow unless they're exposed to photosynthetically useful light.

Buried under a thick mulch layer, seeds receive very little red light, so their phytochrome ratio is skewed to the P_r form. This puts the brakes on any tendency the seeds might have to start germinating, even if water, oxygen, and nutrients are sufficient. So the seeds wait patiently, sometimes for months or even years, until the day when the mulch layer is disturbed enough to let sunlight filter through. Then boom! It's off to the races, with the seedling growing frantically in an effort to get leaves out into the sunlight and start cranking out the carbohydrates before the pantry is bare.

Gardeners have personal experience with this phenomenon when they plant their vegetable gardens. Just look at the instructions on seed packets of lettuce, radishes, or carrots. These tiny seeds should barely be covered by soil, because they need sunlight to break their dormancy. On the other hand, the large seeds of plants like corn, beans, or peas can be buried quite deeply. They have sufficient levels of starchy food reserves to feed the developing seeding, so they're not as influenced by light levels as their photodormant companions.

The photodormancy phenomenon is another good reason not to excessively work up your soil before planting. Not only does this destroy soil structure and hurt the roots of nearby trees and shrubs, it also brings huge numbers of weed seeds to the surface. Soils contain seed banks. Whenever you dig up the soil you are making a big withdrawal from these banks—right into your garden.

Seasonal Dormancy

Many small-seeded plants, both weeds and desirables alike, are also seasonal in their emergence. For instance, many of the weedy grasses and relatives like wild onions are spring annuals, meaning they germinate in the spring and grow into the summer. Others, like the wickedly sticky bedstraw, are winter annuals that germinate in the late summer or autumn and grow through the winter. (Fun fact: the tiny curved hooks found on bedstraw are similar to those on burrs, which were the inspiration for the invention of Velcro.) Seeds of seasonal annual weeds can sit there in full sun and not germinate if the season's not appropriate, even if there's sufficient water and warm temperatures. What keeps them in this holding pattern?

For seasonally dependent photodormant plants, the presence or absence of useful sunlight is not the only trigger controlling

Watering and Dormancy

THE PRACTICE
Decreasing irrigation in autumn to prevent late-season growth and encourage dormancy.

THE SUPPOSED BENEFITS
Lack of water will slow plant growth and force them into dormancy, similar to what occurs in many ornamental bulbs, corms, and tubers.

HOW PLANTS RESPOND
Trees and shrubs begin preparations for winter way back in the summer, when days stated getting shorter after the summer solstice. These changes are mostly internal, though you can see overwintering buds forming on many species. Curtailing water during this important physiological process injures the plants. In fact, insufficient summer and autumn irrigation is one of the reasons for poor bud bloom the following spring. Too little water in autumn can cause early leaf drop, meaning fewer resources for the plant to store over the winter. Fine roots die back and soils become less biologically active without water. Dry soil is also less insulative than moist soil.

**Plants need water all year around.
Don't cut back in autumn.**

Keep your plants and soils well watered into autumn. When you start to see the normal leaf color change, that's your cue that winter dormancy is kicking in. Then you can probably put away the garden hose.

germination. They also need a clue to what time of year it is. As gardeners know, rainfall and temperature can vary wildly throughout the year and even from day to day. Plants need a more reliable system of figuring out when it's time to start growing.

Once again, phytochrome's ability to switch back and forth between forms is what plants depend on. In this case, however, the phytochrome shift is used to measure relative amounts of light and dark during a 24-hour period, what scientists call a photoperiod. Nearly all life forms on Earth have an internal clock or circadian rhythm (*circa* means about, and *dian* refers to day, so a circadian rhythm is about 24 hours). Let's see how plants can use phytochrome shifts to figure out what time of year it is.

I'm not in my garden much on December 21, because it's cold, wet, and, where I live in Seattle, dark for most of the day. This is the winter solstice, the shortest day of the year for those of us in the northern hemisphere. (In the southern hemisphere, June 21 is the shortest day, and for those living on the equator, there's not much difference between the longest and shortest days.) Anyway, back in my Seattle garden, we get about 8 hours of daylight on the winter solstice. By the spring equinox on March 21 the whole world is getting exactly the same amount of light and dark (12 hours of each), and both my plants and I are starting to make appearances in the garden. On June 21, the summer solstice, we have about 16 hours of glorious daylight, and by the autumnal equinox on September 21 we're back to equal day and night. And both my plants and I are thinking about cocooning through the upcoming winter months.

The important thing about this pattern is that it's exactly the same, year after year, and this pattern of changing ratios of light and dark is what phytochrome measures. Remember, the P_{fr} form of phytochrome is changed to the P_r form either by far-red light (a rapid process) or by darkness (a slow process). During the summer, plants have an abundance of P_{fr} phytochrome in their tissues, while in the winter they have more of the P_r form. This ever-changing ratio of P_{fr} to P_r can tell plants exactly what day it is, which allows annual seeds to tell if it's the right season to start germination.

Observant gardeners will point out that not all spring annuals starting growing on exactly the same day, even those of the same species. That's because phytochrome is only part of the alarm clock. The other part of the wake-up process includes warm temperatures, sufficient water, soil chemistry, and lots of other environmental factors that vary from place to place.

Sleeping Beauties

Internal clocks aren't just for waking up, they're also part of going to sleep. For plants, this means going dormant and existing in a state of suspended animation. We've just seen that seed photodormancy is controlled by the absence of red light. Your lawn will stop growing if you forget to water it, but as soon as you add water, off it goes again. Unlike these examples, however, true dormancy is controlled by the internal clock, rather than by some environmental factor like sunlight or water. Dormant plants will sleep, Rip Van Winkle style, until their internal alarm wakes them up.

My own garden is filled with plants that live more than one year. The trees, shrubs, herbaceous perennials, bulbs, and groundcovers all go dormant in autumn. In Seattle and other regions of the world where freezing temperatures are common, dormancy occurs in the winter. Plants in other areas such as deserts and grasslands may go dormant in the summer, when temperatures are high and water is scarce. But as we discussed earlier, rainfall and temperatures are unpredictable: it's not unusual to have warm, sunny days in the winter or to have unexpected rainfall during the dry season. Dormancy allows plants to ignore these unseasonal temptations and remain prudently asleep. Species that weren't dormant would joyfully burst forth, only to be gunned down by the next hard freeze or extended dry period.

Let's look at what happens to my collection of rhododendrons through the seasons. In the spring, the flowers are the first to emerge from their buds—cautiously if it's still cold and more vigorously if we're having a warm spring. As the trusses fade, the leaf buds begin expanding and the year's vertical growth is quickly added through

In spring, rhododendron flowers emerge after a deep winter's sleep. Leaves produce sugars all summer long as next year's buds begin to form. In autumn, the rhododendron is ready for the first cold night to initiate dormancy. Though the plant is dormant, evergreen leaves continue to produce food on sunny winter days.

June. By midsummer, expansion has virtually stopped and the shrubs settle into mass photosynthetic production. I can also watch next year's flower and leaf buds emerge and swell during the long, warm days of summer. By autumn, the bud scales are starting to harden and soon everything is fully protected from harsh winter temperatures. Plant physiologists call this state cold hardiness.

More important to my rhodies' survival is what I can't observe, what's going on inside. Once June 21 has passed by (all too soon!), my shrubs start preparing for winter. The days are already getting shorter, and phytochrome rings the warning bell that winter is on the way. The biochemical processes that get plants ready for winter are complex and time consuming, so they must start long before autumn arrives. When night temperatures begin to dip low enough, the rhododendrons are ready to add the final touches to becoming fully cold hardy.

This same scenario is played out by every tree and shrub in my garden. The timing varies, of course. My rhododendrons have long stopped growing, while my coral bark maple tree continues to push

Potassium to Increase Cold Hardiness

THE PRODUCT
Wood ash or other sources of potassium are added to landscape plants to increase their cold hardiness.

THE SUPPOSED BENEFITS
Potassium strengthens cells walls and displaces cell water, making it more difficult for cells to burst during freezing temperatures.

HOW PLANTS RESPOND
We've already learned that plant cells don't explode under freezing conditions, so we'll discount that part of the explanation. But potassium is involved in water movement across membranes, so what do scientists say? It turns out that extra potassium has no clear role in improving cold hardiness. Although a few studies showed a positive effect, many more found either no effect or negative effects. Once again, adding a nutrient to soil without knowing its existing concentration is a great way to cause mineral imbalances. Get that soil test!

out new leaves almost to the last possible minute. If I lived at a higher elevation, the process would start even earlier, because winter freezes come earlier here than they do in the lowlands. Given enough time, plants become acutely sensitive to their local conditions while always cocking an ear to the ticking of their internal clock. The total time a plant sleeps is usually enough to get it past dangerous freezing conditions. Once that specific period has ended, the plant is ready to begin growing again, and spring bloomers reward us with an ever-changing bouquet of botanical beauty.

To Bloom or Not to Bloom

Just because plants begin growing in the spring doesn't mean they immediately flower as well. Understanding why plants flower when they do was an interesting puzzle for gardeners and plant physiologists alike. Spring bloomers always flower in spring, though sometimes they might have an extra blooming period in autumn. Occasionally my spring-blooming rhododendrons might have a late bloom or two. Other common garden plants, like geraniums and lavender, flower all through the growing season, whereas still others, such as witch hazel and daphne, flower in the winter. For gardeners this is a blessing, because we can fill our gardens with plants that bloom throughout the seasons.

Early researchers had categorized plants as short day or long day plants based on their normal flowering periods. Autumn, winter, and spring bloomers were short day plants, whereas species that flowered in the summer months were long day plants. Plants that didn't seem to give a fig about what season it was bloomed more-or-less continuously and were labeled day neutral plants. (Coincidentally, figs are day neutral plants!) The key factor is day length, and this was very useful information for nurserymen looking to force potted plants into bloom with artificial lighting conditions.

However, researchers continued to poke and prod at the flowering response and soon discovered an uncomfortable truth: it wasn't day length at all that triggered the seasonal flowering response, but rather the dark period—an uninterrupted dark period. Once again, phytochrome was the sensing system plants used to measure the dark period and determine the exact time of year. Flowering progressed (or didn't progress) from there.

What does that mean? Let's look at chrysanthemums, common garden flowers prized for their habit of blooming in autumn and filling in those seasonal color gaps. Originally, mums were categorized as short day plants, and under normal circumstances you would find them blooming as the days get shorter. At the same time, the ratio of P_{fr} to P_r decreases (in other words, there's less P_{fr} and more P_r) in the mum tissues. When the critical phytochrome ratio is reached, flower development begins.

Now suppose you were to go outside every night and shine a bright flashlight on your mums. You would be resetting their internal clock, which would perceive this night interruption as sunrise (even though it would be a very short one). If you were to measure phytochrome in the mums, you would find more P_{fr} and less P_r as a result of that flashlight interruption. Eventually, the P_{fr} would again revert to P_r in the dark, but then the ratio would shift again when the real sunrise occurs. If you continued to interrupt the dark period with light every night, the critical ratio to initiate flowering would never be reached under these endless summer conditions.

Chrysanthemums and other short day flowering plants are more accurately called long night plants. Exactly what constitutes a long night varies from species to species, but in general the uninterrupted dark period must be greater than 12 to 14 hours. In contrast, long day plants should be called short night plants, because they will only flower if the uninterrupted dark period is less than 10 to 12 hours. Bellflowers and carnations, two other common garden plants, are normally summer bloomers. If you did the flashlight experiment on these or other short night plants, you could extend their blooming season by starting it early or prolonging it later. Eventually, however, other environmental conditions such as cold temperatures will bring this runaway flowering to a screeching halt.

Day neutral plants are really night neutral and will flower whenever

Chrysanthemums bloom in autumn.

other environmental conditions are beneficial. Night neutral plants are good all-season performers and may grow well over a broad geographical range. Roses are an excellent example of a night neutral garden plant. Many weed species like dandelions are also night neutral, which partially explains their prolific (and annoying) flowering and seed dispersing abilities. In any case, shining a flashlight on roses or dandelions in the middle of the night won't do much except cause your neighbors to worry about you a little bit.

Interrupting the critical dark period of a plant may disrupt its normal flowering schedule.

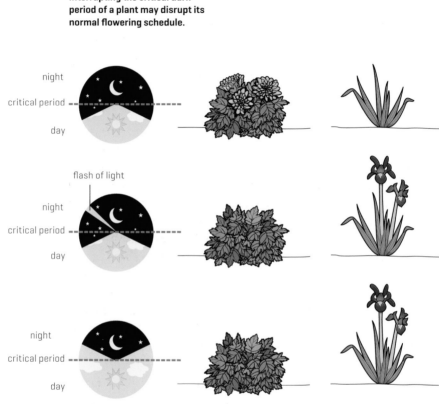

short day plant

long day plant

Sabotaging the Clock

The evolution of shape-shifting phytochrome is an exquisite example of how plants can interpret and react to their environment without sensory organs such as eyes. For eons the phytochrome ratio has successfully informed plants about seasonal shifts so that dormancy and flowering can be carefully controlled. Unfortunately, the early evolutionary process did not take into account how one species—we humans—could alter the light environment through technology and throw a monkey wrench into the physiological machinery.

Let's step away from civilization for a minute and consider natural plant ecosystems. Moonlight and starlight are the only nighttime light interruptions you can observe. Phytochrome machinery is not sensitive to this kind of low-level light, so any light from the night sky has no effect on it shifting forms. Now let's consider what's happened in urban and suburban areas over the last century. Advances in technology have led to the development of artificial light sources, from incandescent light bulbs to fluorescent tubes to high-intensity street lamps. These artificial light sources can cause phytochrome to shift forms, as we saw with the flashlight and chrysanthemum experiment. At your own home, high-powered security lights can affect flowering, dormancy, and cold hardiness of your garden plants. Diffuse, softer lighting, such as the glow from porch lights and solar-powered pathway lights, doesn't seem to have an effect.

This realization can both explain some odd plant behaviors as well as help you avoid damage to your house and garden plants. Let's consider the various holiday cacti that many gardeners nurture indoors or out. Their common names of Thanksgiving cactus, Christmas cactus, and Easter cactus should be big clues that they are all long night plants with slightly different critical ratios of phytochrome needed for flowering around those particular holidays. I have my late grandmother's prize Thanksgiving cactus on a landing in a stairwell with indirect outdoor light. At night, of course, the stairway light is turned on and off several times. And as you might expect, we don't get flower buds on the side of the plant facing the stairs. But the other side of the plant apparently doesn't get enough light from the stairway to

The Season for Tree Planting

THE PRACTICE
Planting new trees in the spring.

THE SUPPOSED BENEFITS
If they are planted after the chance of a hard freeze is past, trees will establish more successfully.

HOW PLANTS RESPOND
Regardless of where you live, spring planting happens right before summer heat and/or drought. Newly planted trees will be establishing new roots for several months after transplant, and if the weather is too hot, the humidity too low, and the soil water too scarce, then the tree will suffer.

It's really better to plant trees in autumn, when the aboveground parts are going dormant and aren't using water. Roots, in contrast, never go dormant. Their growth slows with decreasing temperature, but a protective organic mulch layer will help keep the soil from freezing and roots will continue to grow, using all of the tree's available resources rather than just a portion.

Trees planted in the spring often suffer drought stress during the summer.

Holiday cacti bloom in response to shorter days.

prevent it from setting buds. Once the buds are set, I turn the plant 180 degrees and let the other side of the plant get into flowering mode.

Some plants are a little pickier about their light exposure. Poinsettias are a good example, because absolute darkness is required 12–14 hours per day for several weeks to crank out the flowers. I've had questions from worried commercial growers who were concerned that new ball fields (with night lighting) would cause enough light pollution to affect their poinsettia production. Even worse, a local nursery lost an entire greenhouse of poinsettias when a student entered the greenhouse at night and turned on the lights shortly before the holiday season.

Poinsettias require extensive periods of
uninterrupted dark to bloom.

It's not just flowering that can be foiled by interrupted dark periods. Woody plants preparing for dormancy can be confused by high-
intensity lights, such as those used along streets or in high security areas.
The phytochrome perception that summer has been prolonged can delay
dormancy to the extent that affected trees and shrubs may not be cold
hardy enough to avoid damage of tender tissues by winter freezes.

Native Plants

THE PRACTICE
Using only native plants for landscaping.

THE SUPPOSED BENEFITS
Native species are adapted to the area so they require less water and fertilizer and are more resistant to local diseases and pests.

HOW PLANTS RESPOND
If you and your plants live in a developed part of the world, you aren't really in the original, native environment anymore. Land once dominated by forests or grasslands no longer has much of these original plant communities left. Soils have been stripped and compacted and amended. Urban areas aren't natural, and some of the plants that once grew there won't tolerate these drastically different conditions. Those that do hang on may be more susceptible to opportunistic pests and diseases, rather than more resistant. And stressed natives may well require additional water, fertilizers, and pesticides in their struggle to survive.

Limiting one's garden to strictly native species can decrease your plant diversity, which in turn limits the diversity of birds, insects, and other nifty animals that can enrich your landscape. Instead, make room for some wisely chosen, noninvasive ornamentals that can improve species diversity, provide resources for wildlife, and be aesthetically appealing.

If plants were like animals, they'd probably pack up their petals and move south for the winter. But plants can't move—at least not in a migratory sense. Plant parts, however, can reposition themselves in spite of having no skeleton or muscles. Let's find out how and why sunflowers lift their leaves and clematis swirls their stems.

Night Shifts and Other Unexpected Movements

7

SOME OF MY FAVORITE spring bloomers are tulips—all colors and forms, from the prim species tulips to the drama queen hybrids. My aesthetic side loves the simple yet bold stalks and leaves topped by the perfectly symmetrical flower, and my scientific side loves the fact that every morning they're closed up tight, opening with the sun and closing again at night. This simple garden flower demonstrates, day after day, that plants move.

We're used to thinking about our garden plants as rooted in place, and of course they are. But they can burrow, crawl, twine, capture, and strangle as well. Some plants are lightning quick, whereas others need time-lapse photography to show they've moved at all. These movements may result in permanent changes to a plants appearance or they may be fully reversible. Let's find out how and why plants move.

Tulips open and close in response to sunrise and sunset.

Nasty Plants

My garden tulips and many other plants close up their floral parts and/or leaves at night. This phenomenon is called nyctinasty (from the Greek words for night and pressing close). So the onset of nightfall causes the petals of tulip flowers to press close together. This protects vulnerable tissues from nighttime snackers, and it may help insulate spring bloomers from freezing damage.

How do plants know when it's time to close up shop for the night? We've already explored how phytochrome allows plants to measure time, and this is another example of phytochrome at work. When the sun sets, red light disappears and phytochrome changes into its alternate form, which triggers the response. Sunrise reverses the process, when phytochrome absorbs red light and transforms into its daytime form.

Nyctinasty is easy to observe in the summer, but the cold short days of winter can show us yet another interesting phenomenon. Now, I'm in a long-term relationship with rhododendrons. I grew up in Washington State, where it's the state flower, and I studied their cold hardiness for my doctoral research. I've had rhododendrons at every house I've lived—even in Buffalo! And there's one thing they do that I don't see in other garden plants: they roll up in response to cold temperatures.

Rhododendron leaves are thermonastic. When the temperatures are below freezing, they droop and curl longitudinally, so it looks a little like long green cigars hanging from the branches. Many gardeners think this is a wilting response, but the leaves are full of water during this reversible movement. It's thought that drooping and curling allows the leaves to avoid damage caused by low temperatures, perhaps by creating a warmer cylinder of air inside the furled leaves. Once temperatures rise above freezing, the leaves perk right back up, none the worse for wear.

Gardeners are familiar with both nyctinasty and thermonasty, but even non-gardening types are fascinated with this next nastic movement. People literally cannot keep their hands off sensitive plants (*Mimosa* species), whose delicate leaflets fold up on themselves when

Rhododendron leaves roll
longitudinally when temperatures
dip below freezing.

brushed by a finger. This is thigmonasty, or plant closure caused by touch. The advantage to this is obvious: hungry insects alighting on the leaves are quickly dislodged when their landing strips suddenly disappear.

But the titan of thigmonasty has to be the Venus flytrap (*Dionaea muscipula*). I remember my mother buying one of these when I was about seven years old. It sat on our kitchen windowsill, where I would alternately force-feed it dead insects or play with the leaves. Not surprisingly, it didn't last long, as it probably spent more energy reacting to my pestering than it gained in consuming food.

Plants have other nastic movements, in response to certain chemicals and water, for instance. The relatively rapid response causes many people to think that plants have nervous systems like animals, but of course they don't. Instead, the mechanism behind flower and leaf closure relies on regulating water movement.

Water Works

Let's look at a sensitive plant leaf. It's structured like a feather, and each little green blade is a leaflet. You'll note there's a swollen bump at the base of each leaflet and at the base of the entire leaf. This bump is a pulvinus, derived from the Latin word for pillow. Rather than being filled with feathers, however, the pulvini are plumped up with water.

When a mimosa leaf is brushed, a series of chemical changes is triggered inside each nearby pulvinus, and the result of this chemical cascade is that water rushes into some parts of the pulvini and out of others. The asymmetric turgor causes opposite leaflets to fold together. If there's a great enough disturbance, the pulvinus at the base of the leaf itself expands, causing the entire leaf to fold downward. Essentially, the plant has hidden its tender green parts from insect or other herbivore damage. Once the threat is past, water seeps from the pulvini back into the surrounding tissues, the leaf petiole rises, and the leaflets unfold.

Venus fly traps also close as a result of insect invasion, but rather than avoiding the bugs, fly traps love to get to know their visitors. The blood-red inner surface of each of the trap leaves has three trigger hairs. Curious insects crawling around will eventually trip one of these triggers, but the trap isn't sprung until two of the triggers are brushed within seconds of one another. Immediately the trap closes, the soft spiny projections on the edges of the leaves interlocking with one another like clasping fingers. The insect is held in this photosynthetic prison, which continues to close until the leaf edges, and the insect's doom, are both sealed.

The pulvini of a mimosa leaf are under hydraulic control.

Why does it take two triggering events for the leaves to close? We need to consider the energy that active trap plants expend in closing and opening their leaves. It's always possible that something inedible could trip a trigger hair—some bit of debris blowing in, for instance—and the plant would waste resources closing on a nonexistent meal. So Venus fly traps have evolved this failsafe mechanism to ensure that trap closure will deliver the goods.

Though not as dramatic, the same changes in turgor pressure that capture hapless insects also cause tulip flowers to close at night.

Venus fly trap leaves will only close when two trigger hairs are disturbed in quick succession.

Likewise, our chilly rhododendron leaves droop as both temperature and turgor pressure drop. All nastic movements are rapid, reversible, and nondirectional. In other words, the open-and-close response is always the same regardless of where the stimulation occurred. For instance, you can stroke a sensitive plant leaflet on its upper or lower surface, from tip to base, and the closing response is identical.

Chasing the Sun

Tropisms are another group of plant movements that react differently to environmental stimuli. Tropic (pronounced with a long o) is derived from the Greek word for turn, so tropic movements cause a plant to turn toward (a positive tropism) or away from (a negative tropism) some environmental factor. Some of these tropic movements are quick and reversible like nastic movements, whereas others are slow and irreversible because they trigger a change in the plant's growth.

What gardener doesn't love something about sunflowers? Okay, maybe they're too big and rangy for your yard, or perhaps the yellow-and-black color scheme isn't to your taste, but they're easy to grow, produce prodigious amounts of edible seed, and the young flower heads and leaves track the sun. As sunflowers mature, this sun-worshipping behavior stops as the flower stalks stop elongating and toughen up. They're now stuck in place.

Solar tracking has the lovely formal name of heliotropism. Many garden plants have leaves that follow the sun across the sky. At night, the leaves rotate 180 degrees to face the sun rising in the east. Not many garden flowers actually track the sun, unless you happen to live in an alpine or arctic region. It's in these cold regions where floral heliotropism is most common. By following the sun in these cold climates, the cuplike flowers collect heat, drawing in pollinating insects for a quick warm-up before they depart with their cargo.

In contrast to these sun lovers, plants in decidedly hotter environments often avoid the sun through the very same process. Rather than tilting their flower and leaf surfaces toward the sun, sun avoiders remain parallel to incoming sunlight. These plants are paraheliotropic

Heliotropic plants like sunflowers follow the sun, whereas paraheliotropic species avoid it by keeping their leaves parallel to incoming sunlight.

and they actively reduce the heat load and drought stress that too much absorbed sunlight can bring.

Even though solar tracking depends on sunlight, it's not controlled by phytochrome like nyctinasty is. Instead, solar tracking is controlled by a pigment called cryptochrome (meaning hidden pigment), which absorbs blue light from the sun and uses it as a means to tell time. This pigment sends a signal to the base of the leaf or flower, where a pulvinus works its magic to move the organ to just the right angle. As the sun moves, so do the leaves and flowers, though they stop if clouds block the sun. Once the clouds have passed, movement resumes and actually accelerates until the perfect angle is reached again.

Chasing the Sun, with a Twist

Heliotropism is a fascinating plant phenomenon, but it's not the only trick plants have up their little green sleeves for capturing sunlight. All gardeners know that sun-loving plants hate being stuck in the shade. If they don't languish and die, they try to outgrow their conditions. Many an unwitting homeowner has tried to hog-tie a leaning tree, mistaking the unequal growth pattern for root instability. There's no way to win this wrestling match with nature.

This is phototropism, a directional plant movement toward (or rarely away from) light made permanent by uneven plant growth. Like heliotropism, it's activated by blue light, but the responsible pigment isn't phytochrome or cryptochrome but yet another type of pigment called a phototropin found only in the growing tips of plants. When sun-loving plants desperately seek sunlight in a shady world, that's phototropism at work. You'll also see it in neglected container plants that have fallen over and been left horizontal for too long.

How do plants grow unevenly? Auxins are responsible for cell elongation. If auxins accumulate more on one side of a growing shoot than the other, the plant will begin to curve to one side. In phototropic plants, auxins accumulate on the shadiest side of the growing tips, increasing the growth of the cells on that side, and creating a bend in the shoot. Plant producers take advantage of this directional response, rotating their plants to create intricate spirals (lucky bamboo springs to mind here). Of course, any new growth will resume a normal pattern once the plant is sold by the production nursery.

Phototropic responses are strong in sun-loving plants, but much weaker in shade tolerant plants and most conifers. That's why some trees and shrubs lean away from their neighbors, but others don't seem to mind close quarters. Watch the development of shade in your garden through the years, and selectively prune sun hogs when necessary.

Going to the Dark Side

Most phototropic plants respond positively to light—they grow toward it—but some species grow away from light. As you might expect, roots

**Plants usually grow toward the light,
like this phalaenopsis on a windowsill.**

are negatively phototropic, but why would some shoots grow toward darkness? This fascinating phenomenon is called skototropism (from the Greek word for darkness) and was documented in the tropical forest vine *Monstera gigantea*, whose name should tell you that the leaves on these suckers are huge! In fact, gardeners will recognize *Monstera* species as common greenhouse plants. Needless to say, vines with huge leaves need support, and trees are nature's trellises. Trees also absorb most of the incoming sunlight in the forest canopy, so the darkest areas in a forest will be where tree trunks are stationed.

Now imagine you are a *Monstera* seedling, with a limited amount of resources for survival. You must reach the sunlight before the seed

Directional growth is caused by unequal concentrations of auxin in stems. [A] Auxin is distributed evenly in the shoot tip when the sun is overhead. [B] Auxin molecules shift to the shaded side of the shoot when the sun is at an angle. [C] More auxin causes greater cell elongation on the shaded side, so the shoot bends toward the sun.

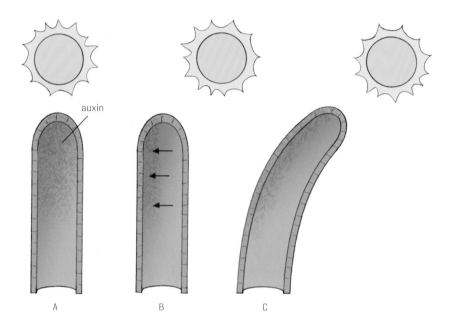

auxin

A   B   C

supply of carbohydrates runs out. You can't see where the nearest tree is, but you can sense the presence of darkness. Growing toward this black hole is the best shot you have for finding a tree. Once a suitable support has been found, skototropic vines become phototropic and reach for the sky like any normal plant.

Skototropism is probably more common than we may realize. Virginia creeper (*Parthenocissus* species) was recently discovered to be skototropic, and there are doubtlessly other vines that temporarily turn to the dark side. The secret behind how vines twine around their support will be revealed later in this chapter.

Staying Grounded

I mentioned earlier that roots are negatively phototropic, but they aren't just repelled by light: they're also attracted to gravitational pull. Gravitropism describes a seedling root's propensity for growing down (positive gravitropism) and the shoot's opposite response (negative gravitropism).

The gravitropic response in roots is variable and depends a lot on the plant's life stage. Seedling roots are strongly gravitropic; they determinedly grow straight down so they can stabilize the seedling and start harvesting water and nutrients. These are taproots, and most plants start life with a deep, strong taproot. But as the seedling develops, it's more important to start colonizing the surrounding soil. New roots develop that are less and less gravitropic and grow horizontally, attracted more to pockets of water and nutrients than to a gravitational pull.

The means by which roots grow downward and shoots grow upward is the same as with phototropism: unequal distribution of auxin along the growing tips of both parts of the plant. For roots, you'll find more auxin in the upper side of the growing tip, and in shoots you'll find more in the lower side. The increased elongation in the first case pushes the growing tip down, whereas in the second it pushes the growing tip up.

But what's really sensing the gravitational field? It's not a pigment, because those sense light. In gravitropism, it's currently thought that tiny grains of starch, called amyloplasts, are the low-tech way that plants figure out which way is down. (It's similar to the way your inner ear senses gravity, though you don't have starch grains in there.) The granules are denser than the cytoplasm, so they sink to the lowest point of the cell. Auxin is allocated to the cells in this same side of shoots, but to the opposite side in roots.

Defying Gravity

By and large, the aboveground parts of plants are negatively gravitropic, as their dependence on sunlight for survival overrules almost any other environmental signal. This dogged tenacity is how lucky bamboo (*Dracaena braunii*) is grown into such interesting shapes. Pots of *Dracaena* are laid horizontally and regularly rotated, creating a spiral or some other design as the plants continue to grow upward.

Gravitropism is also important in horizontal branch stability. As tree branches increase in girth, they get heavier and are more likely to break. Trees form what's called reaction wood, and more of this is formed on the bottom half of the branch than on the top. If you were to look at a cross section of one of these branches, you'd see very narrow growth rings at the top of the branch, progressively becoming more massive as you move to the bottom of the branch. Unequal auxin distribution causes the cambium to expand faster on the lower side, creating this elongated bulls-eye effect.

Plant Associations

When you place a plant in the ground, it changes its surrounding environment. Some of the better things plants might do are provide structural support for climbing plants, create shade for sun-intolerant species, increase soil nutrition through nitrogen-fixing bacteria in their roots, or take up and neutralize salts and toxins in the soil. They don't do it to be nice to their neighbors, they do it as a matter of survival. Unrelated plants in the same vicinity can benefit from these environmental modifications and in turn may provide their own benefits. When both plants benefit, you have what ecologists call a plant association.

Intercropping or polyculture is a gardening strategy that takes advantage of mutually beneficial plant relationships. Together, plant associations increase local biodiversity by attracting and retaining beneficial insects while confusing pests that prefer a monocultural diet. Give a mixed vegetable garden a try! Combine different vegetables with one another to see what works best for your local conditions.

Amyloplasts collected at the base of an upright cell [A] are redistributed when this cell becomes horizontal. The redistribution allows the plant to reorient itself upward [B].

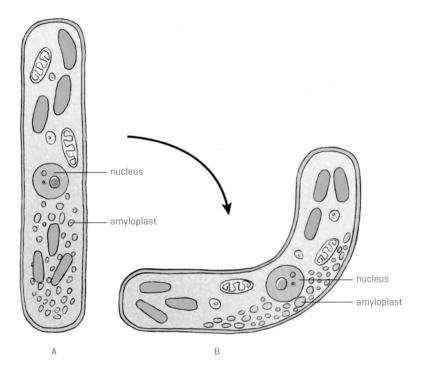

nucleus

amyloplast

nucleus

amyloplast

A

B

Probably the most bizarre gravitropic behavior is found in peanuts. Once the peanut flower is pollinated, it becomes positively gravitropic: it heads for the ground and buries itself to finish development. This is where the mature nuts are found, thus explaining the alternative name of groundnut.

You might wonder whether positive phototropism and negative gravitropism aren't just different names for the same thing. They really aren't. They are sensed by different parts of the plant as a way

Spiraling in lucky bamboo is caused by continually rotating pots under a light source.

of fine-tuning its response to light. For instance, when a sun-loving plant is trying to outgrow shade, it will lean toward light (phototropism), but not at a 90-degree angle. Instead, it continues to grow upward (negative gravitropism) as well as sideways. If you remember your geometry, the hypotenuse of a right triangle is shorter than the sum of the two sides. So the plant is conserving precious resources by taking the shortest distance between two points, rather than stair-step in its efforts to reach the sun.

That's all well and good for woody plants with their built-in structural supports. But what about those vines we left back in the heart of darkness?

Tell Me Why the Ivy Twines

In our backyard we have a wisteria vine that snakes itself around a sturdy trellis. Dotted along the fence line are various clematis cultivars, some of which have lassoed branches of nearby trees and shrubs with their tenacious tendrils. Though the floral show is stunning, the nuisance factor is high, too. I'm perpetually pruning unruly vines out of young trees and shrubs that would otherwise be completely overrun. It's no wonder that some vines, like English ivy and traveler's joy (an oxymoron if there ever was one), have ended up on noxious weed lists.

The winding of vines and tendrils around other plants and upright structures is called thigmotropism. Like thigmonasty in the sensitive plant, physical contact is the stimulus, which in this case causes a directional growth response. Vines have little ability to support upright growth and use thigmotropism to scramble over their neighbors in search of the sun.

My young ginkgo tree is a poster child for clematis catastrophe during the summer. Whenever a young clematis vine would bump into a ginkgo branch, a tendril seemed to emerge overnight and begin to coil. If we were to measure the auxin levels in the tendril, we'd find the levels to be lowest where the tendril touches the ginkgo branch. The high levels of auxin on the opposite side of the tendril cause the cells to grow longer, so the tendril curves around the ginkgo. This unequal growth continues for several rotations until there's a nice little spiral. With the vine firmly anchored, new tendrils will seek and capture ever higher branches.

The curious gardener might wonder whether vines grow clockwise,

counterclockwise, or both. The internet is full of misinformation on the topic, most of it following the popular Coriolis effect mythology, which dictates that vines in the northern hemisphere grow clockwise and those in the southern hemisphere grow counterclockwise. The reality is that more than 90 percent of viny species twine counterclockwise, forming a right-handed spiral. A minority grows clockwise, and a small percentage follows both crowds. Scientists aren't sure what the advantage might be for counterclockwise growth, but observations suggest that the right-handed spiral might be tighter and therefore clingier.

Regardless of which direction they spiral, vigorous vines require serious pruning. So do trees and shrubs, especially those that seem to have their own ideas on how to contribute to your landscape design. What causes plants to grow the way they do?

Garden Care, Not Control

EVER WONDER WHY the Mary of nursery rhyme fame was so contrary, in spite of her obvious success in growing flowers? Maybe it was the trees and shrubs in her landscape that didn't quite act the way she expected them to. Gardeners are often puzzled, if not downright frustrated, when their newly pruned privets or trimmed trees grow back in ways they didn't expect—or maybe not at all! Or they wonder why a gargantuan branch with equally enormous needles suddenly splits the tidy crown of their dwarf Alberta spruce. How about that explosion of bamboo shoots mysteriously appearing in the middle of your lawn? Or those green-leaved shoots appearing on your red-leaved crabapple?

Never fear, this chapter will help you understand why plants grow the way they do. When you're able to think like a plant, you'll be able to predict their moves and won't get checkmated in the garden maintenance game. Let's start with pruning.

Growing Pains: Why Pruning Can Destroy a Plant's Natural Form

We have a lovely noble fir in our front yard. It was barely 6 feet tall when we moved into our house, and now it's close to 20 feet. For the most part, it's been trouble-free: no fertilizers or pesticides needed, just a little additional water in the summer, but otherwise low maintenance. The one glitch we've had is that every other year or so the leader dies back, probably the result of some insect larva chewing away at this young tissue. And every time this happens, we have to do a little corrective pruning so we don't end up with a coniferous candelabrum.

A coniferous candelabrum? Why would the tree put out four or five replacement leaders, rather than just one, like a lizard regrowing its tail? It's one of those intuitions we have about how plants grow that proves to be exactly wrong.

Candelabra trees are prone to breaking.

Whenever an actively growing shoot on a tree is cut, whether by an insect, a browser, or you, it dramatically changes auxin activity. Rapidly growing shoots, either the tree's leader or the tips of branches, garner most of the plant's resources, so the plant can expand its height and width as quickly as possible. To make sure these shoots get the lion's share of the goodies, auxins in the shoot chemically induce dormancy in buds below the growing shoot. When that growing shoot disappears with a clip of the teeth or pruners, the hormonal suppression of nearby buds disappears as well. In a matter of days, you can see these buds swelling and developing into new shoots. However, it's not just one shoot, but several.

All of a sudden, where one shoot was reaching up or out, now four or five new vigorous shoots have taken its place. This can be a desirable characteristic if you're shearing a hedge of boxwood or creating topiary, where bushiness is a blessing. But what about those stately conifers, whose pyramidal Christmas-tree shapes are both natural and aesthetically pleasing? Removing the tops of these trees—not surprisingly called topping (or heading back)—creates a hideous hydra of a tree. Ironically, the reason many conifers are improperly topped is to improve views. Instead, homeowners are left with a flat-top fir with an unruly cowlick that just keeps growing. This means that someone has to continue pruning these new shoots off, every year, to keep the tree unnaturally short and boxy, hardly an aesthetically pleasing view and certainly a long-term maintenance problem.

A Time to Prune

There are times when you *should* prune your woody plants, so keep those shears sharp! You'll need to refer to specific books on pruning, some of which are suggested at the end of this book.

TRAINING YOUNG TREES

Fruit trees need to be trained into a form that allows for vigorous fruit production and ease of harvesting. Likewise, trees and shrubs used for topiary, pleaching, pollarding, and other formal training will need some help in their early years. Be forewarned: trees and shrubs pruned into these artificial forms need to be constantly maintained or they'll try to revert to their normal shape and size.

Formal pruning includes pleaching [LEFT], pollarding and espalier [TOP], and topiary [RIGHT].

MAINTAINING A SINGLE LEADER

When the top of your tree has been damaged or removed, you'll need to keep an eye on the new shoots that spring up to take its place. Select one of these and prune away the competitors within the first year. If the tree is too tall, hire a certified arborist to take care of this for you.

REMOVING THE THREE Ds

Branches that are diseased, damaged, or dead should be pruned from trees and shrubs as soon as they are discovered. With diseased branches, especially, be sure to clean your pruners with a disinfectant such as alcohol or household cleansers. Don't use bleach, because it can corrode your tools (and isn't so great on your plants, either).

RESURRECTING MORTALLY WOUNDED PLANTS

Sometimes there seems to be no way to salvage a damaged tree or shrub. During one winter we got a lot of heavy snow, causing part of our arborvitae to collapse right on top of a nice rhododendron. Most of the rhodie's trunk was broken, and the only thing I could do was cut it off right at ground level. This cure-or-kill approach should only be used if there is no other alternative. It works best with species that sucker easily or naturally develop multiple trunks, and should only be used on well-established plants that have sufficient root resources to help with regrowth. My own rhododendron has grown back quickly and within a few years will reach its original size.

And a Time to Wait

An old adage is to "prune when the blade is sharp," meaning any time you think pruning is needed. For the most part, routine minimal pruning causes little harm. But there are some times when you shouldn't prune at all. A good generic example is pruning conifers back to the wood. Unlike most flowering trees and shrubs, conifers have few dormant buds hanging out under the bark of the trunk and

Cat Faces

Although pollarding can help maintain trees in a smaller than normal form, it also creates unusual rounded structures full of dormant buds. In the spring, the buds break and stems grow in dense profusion from each of the pollard heads. Being a cat person, I always manage to see cat faces in these knobby structures. To maintain this appearance, the branches must be cut back to the

heads every autumn. The heads should never be removed, as it not only destroys the pruning form but will lead to very ugly branch development. Pollarding should only be attempted if you know what you are doing and you're willing to do it for the lifespan of the tree.

Pollarded trees (LEFT) **have knobs of densely concentrated buds** (BELOW).

Rhododendron flower buds remain dormant through autumn and winter.

branches. How many juniper hedges have you seen that have been sheared back to the wood? Looks horrible, doesn't it? And it never, ever fills in. This characteristic of conifers is one of the best arguments I can give for using a flowering shrub species for your hedge: it will tolerate the shearing and come back with a flush of growth from its numerous latent buds.

Speaking of flowering shrubs and trees: don't prune their flowers off! That may seem like silly advice, but it happens every year when gardeners prune their rhododendrons and lilacs and other spring bloomers in autumn. These species and many others set their flower buds in the summer, and the buds overwinter until spring. To avoid losing the gorgeous floral show, prune these species immediately after flowering, and then leave them alone.

Likewise, don't be tempted to help deciduous trees and shrubs with their autumn leaf drop by pruning their branches at the same time. Leaves that turn red in autumn are busy transporting sugars and other resources to the branches, trunk, and roots. Lopping off limbs while resources are being relocated is a sure way to decrease your plant's winter larder.

Another common mistake gardeners make is crown pruning trees and shrubs after transplanting them. It appeals to our sense of symmetry that if we've cut or damaged roots in the process of planting, then we should reduce the crown in the same manner. Conventional wisdom says that fewer leaves means less water needed for the crown, so less stress is imposed on the root system.

That's the way we would expect it to work, anyway. But plants have their own way of responding to crown damage, which is to immediately send out more shoots and leaves to replace those that we've cut off. Anyone who's ever cut back a rose or sheared a boxwood knows this, so it's amazing that we can ignore that experience and assume new transplants that have also been pruned will patiently wait for their roots to grow before they push out new leaves. By crown-pruning new transplants, you remove part of the photosynthetic machinery needed to provide food for establishing roots, and the existing resources are now directed to new leaf growth instead. New leaves mean more water needed from the roots. It's the worst possible scenario.

This is where your new-found knowledge on how plants work will allow you to trust in what you can't see. When plants are putting resources into root growth, they may not grow at all on top. This is especially true of trees that have been root pruned to correct root flaws. You can gently wiggle your new transplant once a week by grasping its trunk and feeling how much it moves. As time goes on, it will move less and less as roots become more numerous and hold the tree more firmly in place. This will happen quickly with bare-rooted plants. The crown will stay dormant until enough roots have established to take up water needed for supporting new leaf growth. At that time, the buds will burst open, birds will sing, and you'll know that your tree is alive.

Color changes indicate that
programmed cell death is occurring
in these damaged salal leaves.

Nature's Little Pruners

We've talked about gardeners as pruners, but what about all the creatures to whom our gardens and landscape are a giant smorgasbord? Insects and other grazers nibble, notch, chew, suck, drill, and mine nutrient-rich leaves, and a plant's response to biological pruning is often different than that to simple mechanical shearing. When a leaf is attacked by an herbivore, there's a flurry of biochemical activity in

Antitranspirants

THE PRODUCT
Acrylic, latex, or wax materials that are sprayed onto leaves to form a barrier over the surface. There are also antitranspirants that cause leaf stomata to close.

THE SUPPOSED BENEFITS
By sealing the surface of the leaf or forcing stomata to close, evaporation is reduced.

HOW PLANTS RESPOND
While antitranspirants do reduce water loss from leaves, they interfere with normal plant physiology. Reducing water loss from the leaves means that water transport and nutrient uptake are reduced as well, because it's evaporation from the leaf that pulls water and any dissolved substances through the plant. The lack of water flowing through the plant also eliminates evaporative cooling of the leaves, making them susceptible to solar heat damage. Finally, closed stomata don't take up carbon dioxide, so photosynthesis is reduced.

Antitranspirants are best used for reducing water loss in Christmas trees and cut flowers. They should not be used on living plants.

By clogging stomata, antitranspirants interfere with a plant's ability to take up water.

the wounded tissue, called the hypersensitive response. A combination of factors, including the physical destruction of leaf tissue, the presence of proteins or hitchhiking viruses in the nibbler's saliva, and exposure to the environment, trigger this response, which has the end goal of isolating the injured parts of the leaf from the rest of the plant. Destroyed tissues are walled off in this process, which has the oddly robotic moniker of programmed cell death. We see these cellular changes as red, yellow, brown, or blackened areas on the leaf, due both to anthocyanin pigments and to reactions between biochemicals that once were separated and now are all mixed together.

Besides the interesting color changes you might see, leaves can be inadvertently infected with viruses or microbes carried by hungry herbivores. Aphids and other insects with piercing mouthparts can inject viruses that create white or yellow mottling patterns on leaves. While viruses are rarely harmful enough to kill plants outright, they are a nuisance and can reduce the vigor as well as the aesthetic quality of the affected plant.

High Stakes

Since we're discussing the proper care of trees and shrubs, let's consider another error some gardeners make when planting trees: staking. Like a stake through the heart will kill a vampire, a misplaced stake will just as surely kill the tree it was meant to support. This mistake, like many others gardeners tend to make, results from not understanding how plants work. Let's look at what happens in nature as a predictor of bad staking.

We're in the middle of a forest. The trees here are tall, skinny, and have few lower branches. All the growth is happening in the canopy as trees compete for sunlight. As we walk to the edge of the forest, we notice that the trees get shorter and stockier; their trunks are decidedly thicker than similarly aged trees in the middle. These edge trees have developed thicker trunks and are shorter in stature as a result of buffeting by the wind. Those in the middle are untouched by the wind and their resources have gone into growing taller, not wider.

Overstaked trees won't develop normal growth forms.

When we stake a tree too tightly, we are literally providing a crutch for the tree by eliminating the effects of wind. Like its wild cousins in the middle of a forest, the tree can put its resources into growing taller, and these trees characteristically look like lollipops with their skinny trunks and lush crowns. There's little trunk girth development, because the stake makes it unnecessary. What do you suppose will happen when (and if) that stake is finally removed?

It's not too difficult to predict that these tightly staked trees will no longer be such upright citizens once they're removed from bondage. The heavy crowns perched on top of tiny trunks cause the entire tree to bend over, if not break outright. There is no way to easily fix this problem, other than by restaking the tree properly and allowing it to develop adequate girth.

Instead, let nature help prepare your trees for life on their own. If they need to be staked (and many do not), stake them low and loose. Be sure to remove the stakes after a year, because anything that can't stand on its own after a year in the ground will never develop that ability. It's better to dig it up and try again, rather than keep the poor thing propped up in a botanical life support system.

Don't Add Insult to Injury

One of the biggest misperceptions people have about plants is that they heal when they are wounded. Yes, plants do recover from damage inflicted by insects, browsing deer, pruner-crazed gardeners, and errant automobiles, but they don't heal like humans do. So treating wounds like you might treat your own scratches and scrapes can be a very bad idea.

We're used to seeing cuts disappear as scabs and then new skin slowly cover and replace them. But this doesn't happen with plants. Once their tissue is gone, it's gone. It can't be regenerated. Instead, the plant creates a wound tissue that both protects the cut area and often has antibiotic activity of its own. In trees, you can see wound wood quite clearly. It looks different than the surrounding bark—it's smooth and often lighter in color—and it swells along the perimeter

Tree Butts

If you've spent any time around older trees, you'll have seen tumor-like protrusions on their trunks. Sometime these growths are closer to eye level, but often they're at the base of the tree, providing the tree with a substantial derriere. These odd deformities are burls, woody growths on the trunks and branches that develop after environmental damage from pests, disease, and even people. This is a different response than the formation of wound wood. Sometimes injury will turn on phytohormones, which results in uncontrolled growth. If the underlying wood also includes meristematic tissue (points of potential growth), then the developing burl can contain numerous dormant buds.

Many trees, both coniferous and broadleaf, can create large burls, and these in turn are prized by woodworkers. However, this should

not be taken as an invitation to remove them from the tree. Burls are thought to be benign and removing them will both injure the tree and remove a substantial supply of stored carbohydrate. Instead, think of these odd growths as badges of life experience. And while some burls develop naturally, gardeners can prevent those caused by mechanical injury. So, be careful with cars, wheelbarrows, weed whackers, and anything else that could physically damage the bark and underlying tissues of trunks and branches.

Burls are unusual but harmless growths on tree trunks.

Wound wood seals and protects cut surfaces.

These suckers originated from roots of one poplar tree.

of the injury. This donut-like structure continues to expand over the cut surface of the wood and ideally will join in the center to create a physical barrier.

Let's consider the bark of my young styrax tree that's perpetually scraped up by squirrels running up and down on the way to the nearby birdfeeder. When the squirrels' claws tear through the bark, nearby cells are called up like paramedics to come to the rescue. These cells begin to form all kinds of defensive compounds to keep the injury and possibly disease organisms from spreading to healthy tissues. In trees, we call this wound wood, and it's quite different than the original tissue. We can see this wound wood on some of the older scratches on the trunk; the texture is different and it will never look like the original bark. These new tissues seal off the wound from the sides and from below, isolating it and any unwelcome hitchhikers from the rest of the tree.

When well-meaning gardeners paint wounds with sealer or cover the wound with so-called natural tree healing products, they interfere with the tree's own ability to seal itself. In fact, oxygen is an important component of the biochemical reactions that are taking place, and these tree wound products effectively keep oxygen away from the very tissues where it's needed. Worse, some of these products are desirable food sources for microbes, which means you've just issued an invitation to invaders. The only thing gardeners should do in treating tree wounds is to carefully trim away any ragged or torn bark edges using sterilized pruning tools.

The only time wound dressings are warranted is if your area is threatened by specific diseases (such as oak wilt) or insect infestations (like emerald ash borer). In such cases, the cuts should be treated with the appropriate fungicide or insecticide. Hire a friendly certified arborist to do this for you rather than dealing with it yourself.

The Amazing Green Machine Replicates Itself as If by Magic!

Back to that bamboo-in-the-lawn puzzle at the beginning of this chapter. Did it come from seed? Not very likely, as most ornamental bamboos don't grow easily from seed, especially in an established lawn. Let's take a peek over the fence at the neighbor's yard. Sure enough, there's a nice stand of bamboo just waiting for the local panda to drop by for a snack. What you can't see is the system of vigorous underground rhizomes that can run for many feet before snorkeling up through the soil surface. If there's enough light in this spot, a new bamboo clump will form. And if you don't take care of it, soon you too will have a panda paradise in your yard.

Many weedy plants spread this way. It's a characteristic that can be annoying, but gardeners can take advantage of it. When you're planting your perennials, groundcovers, and other permanent plants, do a little research to find out how quickly they spread. You might be able to get away with purchasing far fewer plants and just letting them do their thing. I planted five little native strawberry plants as a groundcover about ten years ago. Today my strawberries cover about 400 square feet of the low-maintenance landscape along the front of our house.

But be sure to consider the other side of the coin, too, before you commit to buying aggressively spreading landscape plants. Will you still be happy with that ivy groundcover when it spreads to infinity and beyond? Or are you prepared to care for a thicket of vine maples or willows that has replaced the single specimen you planted a few years ago? Any time you see a nursery tag that says "spreads easily," "naturalizes," or "engulfs anything that doesn't move," you should consider the long-term effect of those deceptively small specimens.

Sprouts, Suckers, and Stress

Have you ever seen what look like wild hairs growing in and at the base of landscape trees? They're ramrod straight, vigorous, and look more like saplings than anything else. These are suckers. The ones in the crown of the tree are called watersprouts, and you can frequently find them in dogwoods, cherries, and other common ornamental species. In general, watersprouts and suckers are structures that trees produce to reinvigorate themselves. While some species naturally sucker as a way to self-propagate, in others suckering can be a signal of stress. Understanding why suckers and watersprouts appear can help gardeners decide how, and whether, to treat the phenomenon.

Watersprouts usually appear when something has happened to the crown of the tree. Perhaps it's been pruned back improperly or a branch has broken out. Regardless of the reason, the plant's response is to create a new crown. Watersprouts can be a great diagnostic tool, though by the time you see them it may be too late to solve the problem. Here's a dramatic example.

Uphill from my place in Poverty Flats is an expensive view home (big house, small yard) that has a single landscape tree. It's an ornamental cherry, quite mature. A few years ago we noticed large watersprouts on the lower two limbs. The rest of the tree was normal—or so we initially thought. It turns out that the upper crown was slowly being girdled by neglected staking wire, which of course meant less food for the roots. It's a shorter distance for water to travel from the roots to lower branches than to higher ones, so the tree was essentially bonsai-ing itself. An impaired root system can service a smaller, lower-crowned tree better than a full-sized one. And, in fact, a few years later this tree did fail, although inexplicably the homeowners have left the butchered remains for all to see. It's a great teaching tool, but not a very good landscape design element.

There was nothing to be done in the Case of the Choking Cherry. Had these watersprouts been observed and their cause diagnosed early on, the tree probably could have been saved. Sometimes trees will send up watersprouts for no apparent reason—we have a dogwood that seems to sprout if I look at it sideways—but generally something has happened in the past that the tree is responding to.

The causes behind suckering are a little more complicated, but still easily diagnosed once you know the root causes. Let's consider that red-leaved crabapple I mentioned at the beginning of the chapter. Many landscape plants have been bred for red leaves, but they often don't have cold hardy rootstocks. So the red-leaved cultivars are grafted onto native rootstocks that are adapted to colder temperatures. In a successful graft union, the vascular tissues of the scion (the top part of the grafted plant) and those of the rootstock fuse and create a functional plant.

Now we have to look at the ultimate flaw in this practice. The native rootstock is adapted to the conditions where this tree is planted; the scion, probably not. Under the best circumstances, the rootstock is kept under control, servicing the scion but pretty much limited to life

Watersprouts are vigorous upright branches.

downstairs, as it were. But sometimes rootstocks revolt and send up their own shoots. The native species are green-leaved and vigorous, usually able to outgrow the red-leafed scion. It's crucial for gardeners to remove these suckers as quickly as they appear or the rootstock will literally drain the nutrients away from the scion and eventually subsume it. The same is true for any grafted specimen, where the scion has been chosen for an unusual leaf color, crown architecture, or some other horticultural curiosity that renders it less vigorous than the rootstock. Working classes arise, indeed!

Sprouts on Steroids

Many of the plants we cultivate for our gardens are mutants, with their arthritically twisted limbs, leaves splattered with white blotches, and forms of every shape and size. In the real world, these botanical oddities wouldn't survive long because their mutations make them less likely to reproduce successfully. Yet even in our protective care, you'll sometimes see an odd branch appear in the middle of a cultivated tree or shrub that is definitely not like the others. In dwarf species, like the Alberta spruce I mentioned earlier, this branch may be enormous in comparison. In variegated species, you might have branches with all-green leaves. These are reversions to the wild type, and they are always more vigorous than the cultivar. For this reason, you'll want to cut these sports out as soon as you see them, because they'll suck the lion's share of resources from the rest of your tree. Nature always finds a way to overcome whatever constraints we put on it.

One Big Happy Family

The easiest way to kill a tree, slowly but surely, is to girdle its trunk. Girdling can happen accidentally, as when string trimmers and lawn mowers get a little too close for comfort. Or it can be a deliberate and often illegal act by disgruntled homeowners seeking to improve their views of nature by removing all vestiges of anything green. In either case, the complete removal of a band of bark and the thin layer of live

tissue underneath effectively shuts down the transport of sugars from the leaves to the roots. The roots are alive, but they literally starve to death, because they can't make their own food. Through this extended dying process, the roots continue to take up water to support the crown, but as their energy reserves dwindle they sputter to a stop. We finally see the results of this when the leaves wilt from lack of water. Mercifully, the end comes pretty quickly after this point.

On occasion, though, girdled trees appear to defy death and continue

to grow happily, even without a continuous band of phloem running through the trunk to the roots. It's most common to see this in groves of alders, willows, or poplars in wetlands near beaver dams. The beavers girdle trees in part to feed on the nutritious tissues underneath and, in the case of smaller trees, to create lumber for their building projects. So how do these lucky trees appear to transcend mortality?

It's a combination of root physiology and genetic compatibility that saves these trees from becoming part of Chez Beaver. Underground, plant roots grow in all directions, frequently crossing over one another. If the roots belong to trees of the same or a closely related species, they will fuse or graft together as they thicken, effectively linking the two trees in a permanent, intimate relationship. They share water, nutrients, hormones, and anything else that can be carried through the xylem or the phloem. The partnership can be extended through many individuals, so that an entire forest of the same type of tree is joined by one vast, interconnected root system. When one member of the cooperative is damaged by girdling or some other environmental injury, its roots essentially become dependent on the neighbors for food. The aboveground portion of the tree remains alive and functional, and the entire grove benefits because competing species can't grab a foothold, an obvious risk if a tree were to die, fall, and create an empty space waiting to be filled.

Root grafting is also important for gardeners to understand when they apply translocatable herbicides. These are herbicides, like Roundup, that move through the plant courtesy of the phloem and kill the entire plant, roots and all. Plants that are connected by roots, whether they are grafted individuals or clones along a stolon, will all be affected by translocatable herbicide applied to just one of their members. This can be a boon to gardeners hoping to take out a stubborn cluster of hedge bindweed, for example: you only need to spray a few of them to kill the entire group. On the other hand, there are unhappy, unintended consequences. Many a gardener has innovatively sprayed the annoying sprouts arising from a lilac or some other suckering species, only to find that the cherished original tree is also deleted from the landscape.

Finally, we need to remember the role our mycorrhizal fungal partners play in the underground railroad. Though it's clear that water and nutrients can be shared between the fungi and plants using mycorrhizal connections, it's not clear whether herbicides would move through the pathways. In fact, some mycorrhizal fungi are able to degrade herbicides and use them as a food source. This fascinating science continues to evolve, and for now gardeners can only marvel at the interconnectedness of life underground.

**Root fusion may occur
underground or aboveground.**

Plants are masters of the ground game, with solitary individuals able
to colonize huge swaths of land through cloning. The clone war might
be a good strategy in such cases, but it creates a monoculture that can
be wiped out by disease. It's much better to do a little genetic rear-
rangement with another plant partner. But how to find that special
someone when you're literally a homebody?

Finding Love in a Sedentary World

9 **SEX IS COMPLICATED,** and it's even more complicated for plants. Unlike people, they can't go looking for Mr. or Ms. Wonderful with whom to start a family. Nor are their offspring able to pick up and move out on their own. Like other sedentary species, plants rely on the environment or living creatures to help out with matchmaking and house hunting. But plants have met this challenge in astonishing ways, often using unwitting animals, including gardeners, as a means of their own reproduction, dispersal, and survival. And they've been doing it for millions of years. While we can't travel back in time to see how this started, we can look at some descendants of prehistoric plants.

In the Beginning

Woodland garden staples like mosses and ferns are ancient plants and structurally pretty simple compared to more recent botanical arrivals. Even so, they've obviously survived quite nicely despite their primitive nature.

As ancient as ferns are, they remain a popular plant choice for many gardeners, including me. Most of our ferns have volunteered to be part of our landscape, popping up along fence lines and among established shrubs. They arrived as microscopic spores, blown in on the wind or carried by raindrops and lucky to have landed where they found enough water to begin life. If you turn over the fronds of most ferns, you can find little brown bumps on the underside. These intricate patterns are spore nurseries, called sori. Mosses have tiny capsules containing spores; the hated horsetail has an asparagus-like structure containing spores at the end. (It's this toxic stalk that's sometimes mistaken for real asparagus by unfortunate foragers.)

Liverworts produce genetic clones [TOP] as well as genetically unique spores shed from tiny umbrellas [BOTTOM].

And liverworts (*Marchantia* species), one of my favorite little plants that's equally loathed by nurseries, open tiny umbrellas from which spores are spread. Even cooler, the leaves of liverworts have miniature cups filled with clones (with the lovely name of gemmae) that sail off to new homes whenever dislodged by a raindrop.

Liverworts provide a good example of the two ways that plants can reproduce: asexual clones and sexually produced spores (or seeds in most plants). While most plants can do both, some like orchids can only reproduce sexually. What's the benefit of having more than one way to reproduce? After all, most animals reproduce sexually and only the more primitive members clone themselves.

Botanical Xerox Machines

Let's consider cloning, or vegetative reproduction, first. It's a fast and easy way of establishing territory, especially in environments where individuals might be isolated from one another. This is the strategy that many invasive species use: the mother plant provides hordes of storm troopers to conquer new environments. The clones are genetically identical to the mother plant and to each other, and given that an invading plant has been tough enough to survive and establish a new outpost, its clones will likely also be tough survivors. We use this characteristic in our own gardens, when we plant groundcovers. We don't buy hundreds of plants, but just a few with the knowledge that they will spread and merge. By understanding the ways that plants spread vegetatively, we can use this information to propagate plants ourselves as well as predict how well new plants will spread in our garden before we plant them.

For millennia gardeners have shared their plants with other gardeners, and plants make it easy with their remarkable cloning ability. Many species simply make carbon copies of themselves; we call these multiple crown species or sometimes multi-trunked in the case of trees and shrubs. Whichever name you prefer, these species are easily divided by cutting the connecting roots and separating the crowns from one another, then replanting. Multiple crown species can create vast thickets of themselves. The largest known is a single quaking aspen clone spread over 100 acres in Utah.

In contrast, probably the most unusual method of cloning is leaf propagation. When most plants drop their leaves, these tissues die. But in some species the leaves develop roots on the detached end. If you have ever had a jade plant, you'll have noticed that fallen leaves around the plant often root themselves. Other species, including African violets, snake plants, and begonias, send up new plantlets from leaf ends and cut veins. Kalanchoe, sometimes called maternity plants, form tiny plantlets along the margins of attached leaves; these fall to the ground and take root. It seems that most plants that successfully clone from leaves are succulent species or rainforest plants. In both cases, there is plenty of water available for the rooting process, either contained in the succulent leaf or in the environment itself.

We're more used to seeing plants rooting from stems, either in the garden or as cuttings on our kitchen windowsills. Stems have a Tinkertoy type of structure consisting of nodes and internodes. The internodes (literally, places between nodes) are really nothing more than straws connecting nodes to one another. The nodes are wondrous little places containing meristematic tissue from which all kinds of new things can arise: new leaves, new roots, even new plants. This is why it's important for gardeners to include nodes on their cuttings; without the node, all you have is a straw. (And because there's no "this end up" sign on a stem, be sure to make the bottom cut at a 45-degree angle so you don't root it upside down.)

The majority of my landscape shrubs have more or less upright, woody stems. But as the weight on these stems increases, they're often bent toward the ground, where they can take root. Azaleas, spirea, honeysuckle, and many other ornamentals have this ability, and given enough time these rooted branches can survive on their own if they become separated from the original plant. Yet another desirable botanical trait for the frugal gardener to take advantage of!

But other plants in my landscape, especially groundcovers, have a different stem arrangement. The bright red runners of my native strawberry that crisscross the soil are horizontal stems, or stolons,

When leaf-borne plantlets fall to the ground, they are ready to take root.

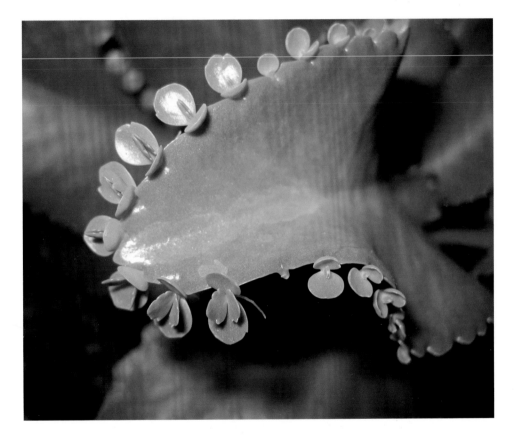

that begin within an established plant. Along the runner, like beads on a string, are little plantlets that develop roots and start to grow if they happen to land on soil. The entire system is connected, so that roots from established plants provide water and nutrients to the runners as well as the developing, but unrooted, plantlets. Spider plants also have

Mixed groundcovers hide a maze of rhizomes and runners underneath.

stolons. When planted in hanging baskets, the effect of all the little plantlets dangling from the stolons really does look like spiders.

In contrast to my sun-loving, running strawberries, my landscape also has Oregon oxalis, a lovely understory native that pops up here and there as a result of a different stolon system. Oregon oxalis has underground stems called rhizomes, which also crisscross the landscape, though in a more secretive way. Only by moving aside the soil can one see the rhizomes and discover that their structure is much like that of runners. Many plants, including some truly annoying weedy species, spread by rhizomes.

While runners and rhizomes are recognizable as stems, other plant parts used for vegetative reproduction look more like roots. These stems in disguise include bulbs, corms, and tubers.

BULBS

Bulbs are immediately recognizable to even the non-gardener, as everyone has peeled an onion at some point. Those layers of scales are actually modified leaves. So, the onion is just a flattened, white stem with fleshy white leaves that surround, protect, and nourish the bud at the center.

I've planted (and forgotten about) so many bulbs throughout the seasons in my garden that I'm assured of surprises all year round. Where only one flower might appear the first year, in subsequent years there are more and more blooms. Bulbs vegetatively reproduce by forming offsets, tiny bulblets nestled in between the scale leaves and the central axis. The offsets of some species, like bluebells, can flower the next year; others, like tulips, may need to develop for a few years before flowering. The bigger the bulb, the more food is stored in its leaves, and the bigger the flowers will be.

For this reason, many gardeners dig up their dormant bulbs every few years, removing the offsets and replanting them elsewhere so they don't compete with each other. Digging animals do the same when they disturb the soil. Whether it's a gardener or a squirrel, both are used by bulb-forming plants as the means to spread their clones throughout the landscape.

Contractile Roots

I'm a lazy gardener, so I tend to choose plants that take care of themselves. Thus, I use a lot of bulbs and corms in my garden. When I buy bags of tulips, daffodils, or crocus, I dutifully plant them to the proper depth, and then ignore them. They grow happily, bloom profusely, and spread prolifically. It's easy to understand how naturally dividing bulbs can stay at the proper depth in the soil. But what about the seeds that drop to the ground, germinate, and plant themselves at the proper depth?

You might assume that animals bury them, either by accident or to store as food, but that's a pretty unreliable way to ensure survival. Instead, the developing seedling eventually forms a tiny bulb (or corm) with specialized roots that actually pull the bulb into the soil. If you look at these roots closely, you'll see that they look wrinkled. In fact, they look very much like a miniature version of the popular shrinking garden hoses. These contractile roots expand with water and then contract, pulling the bulb deeper and preventing it from becoming dislodged, damaged, or eaten.

It's a great protective mechanism for ornamental bulbs and corms. Unfortunately, it works just as effectively with annoying garden weeds like wild garlic and yellow nutsedge.

contractile root

Contractile roots protect bulbs by pulling them deeper into the soil.

Bulbs [TOP LEFT], tubers [TOP RIGHT], and corms [BOTTOM] are botanically different, but all of them can create new plants.

CORMS AND TUBERS

Like bulbs, corms have a flattened stem where roots emerge from the lower surface, but on top they have a rounded lump of storage tissue rather than neat, concentric layers of leaves. Garden favorites like crocus and gladiolus grow from corms, and their offsets (or cormels) can be found circling the bottom edge of the mother corm.

Tubers have no obvious base and instead have a swollen, lumpish appearance. They look like potatoes. And, of course, a potato is a tuber. Most of the tuber is storage tissue, except for the eyes, which are dormant buds. When the buds begin to develop, new plantlets are formed using the tuber's stored reserves for food and water. So when you find that neglected potato that fell behind the shelf, you'll understand that it's so shriveled and flaccid because it's supporting all that new, lush shoot growth. Just cut up that old potato and plant it!

Let's Talk about Sex

Given all the benefits of vegetative reproduction and the myriad ways that plants can clone themselves, why do plants go through the fuss and bother of sexual reproduction? Well, anyone who enjoys breeding roses, tomatoes, or just about any other plant can tell you the answer to this one: it's the way to create new cultivars. From the plant's perspective, it's strictly a matter of survival. By mixing things up, genetically speaking, plants can ensure that at least some of their offspring will survive in an ever-changing world.

Let's look at one of my favorite plants, gardenias. I have tried for years and years to grow gardenias successfully as a houseplant. No matter what I did—no matter which window I chose or which fertilizer I used—I could never get the masses of pure white blossoms with their exotic fragrance. Sure, the plant would live, but it just wouldn't flower. And, frankly, gardenias without flowers are pretty boring houseplants.

But several years ago a new gardenia called 'Klehm's Hardy' became available. This cultivar can withstand moderate freezing, so now I'm able to grow my favorite fragrance flower outside. Through careful plant crosses, gardenia breeders were able to create a cultivar that can withstand lower temperatures than gardenias would experience in their native tropical and subtropical environments.

Without sexual reproduction, plants wouldn't be able to develop

resistance to insects, disease, and a whole range of environmental stresses. Plants tend to be rather lax in their selectivity of sexual partners and can be found happily hybridizing with other species, and sometimes even other genera. Many of these hybrids are aesthetically pleasing, resistant to stress, or have other characteristics that we enjoy. The London plane tree (*Platanus ×acerifolia*) is a hybrid between the Oriental plane tree (*Platanus orientalis*) and the American sycamore (*Platanus occidentalis*). This hybrid's ability to withstand tough urban conditions and provide ample shade makes it a popular street tree. Crosses between raspberry and blackberry, plums and apricots, and mandarin oranges and grapefruits have given rise to delicious loganberries, pluots, and tangelos, respectively. And most of our favorite flowers are hybrid crosses, courtesy of nature and nursery alike.

Why They're Called Shrinking Violets

A person described as a shrinking violet is pictured as shy and retiring. This is hardly an apt description of real violets, which are such prodigious seed producers that many gardeners consider them weeds. It's likely the name arose after observant gardeners noticed that violets not only display their small, showy five-petaled flowers, but bear hidden flowers as well. Found close to and sometime beneath the soil surface, these flowers never open: they are cleistogamous, which translates colorfully to closed marriage. As the name implies, the purpose of cleistogamous flowers

is to produce self-pollinated, fertile seeds, which quickly fill in gaps around the parent plants and create dense patches of violets.

The phenomenon isn't limited to violets, either. Several other well-known garden plants have this interesting adaptation, including many grass species and legumes such as peas, beans, and peanuts. Cleistogamy is a common survival strategy for plants in harsh environments. They're always able to reproduce even when they're all by their lonesome. And now you have a fun new word (klī-stŏg'-ă-mē) to trot out at your next social function.

Some violet flowers can only be seen by digging up the plant.

We know that animals sometimes hybridize: ligers from lions and tigers and mules from horses and donkeys, for example. But these hybrids are usually unable to reproduce, and therefore they're just an evolutionary dead end. Hybrid plants, on the other hand, are often fertile and can reproduce themselves. This makes the family trees of many garden favorites, like rhododendrons, infernally complicated. The parentage of one of the rhododendrons I used for my doctoral research consisted of one species, several cultivars, and a helpful dose of unknown.

Not all hybrids are desirable, at least as far as gardeners are concerned. Allowing certain squash varieties to cross-pollinate, for example, can result in some truly vile produce. Many gardeners who grow heirloom vegetables and collect the seeds go to great lengths to avoid exposing their garden gene pool to undesirable outside influences, like a shotgun-toting father might protect his teenage daughters. Uncontrolled plant hybridization can result in muddy flower colors, lack of fragrance, and other characteristics that gardeners find unappealing. But the purpose of plant promiscuity is to constantly produce new combinations of genes that might help the next generation survive future challenges—not to please people. When hybrids do both, however, they ensure their future survival by using gardeners as the means to that end.

THE BIRDS AND THE BEES

We've discussed the benefits of sexual reproduction that cloning can't confer to plants, but we still haven't mentioned how it works. It's time for the talk.

Birds and bees, bats and butterflies, wind and water: plants use all of these as transport systems for pollination. Conifers, the oldest seed-producing plants, set their pollen on the wind in search of receptive female cones. In fact, any plant that produces clouds of yellow pollen, like pine trees, grasses, and ragweed, belongs to a wind-pollinated species. In grasslands, dry pine forests, and other environments where species diversity is low but plant numbers are high, wind is a low-cost method of sperm delivery.

In more complicated ecosystems, where many species jostle for space and individuals of the same species may be few and far between, other methods of pollination are needed. This is where plants use their flowers to entice animals to become willing—if unwitting—delivery services. As we'll see, it's a more expensive process for the plant, but it increases the likelihood that pollen is delivered to another flower of the same species.

For animals to willingly act as plant matchmakers, they need to receive some kind of payment. Plants reward their pollinators by providing food (nectar and pollen), nest building materials (oils and waxes), heat, or other necessities. To attract pollinators, flowers use color and odor as advertisements, and these odors and colors are often fine-tuned for specific pollinators. When we gardeners understand how birds, bees, butterflies, and other pollinators see the world, this can help us select flowers that will attract these garden visitors.

I have a lovely ruby-red glass hummingbird feeder hanging in my south-facing garden, and hummingbird lovers know that red is a color easily seen by birds. Fuchsias, columbine, hibiscus, and other red and orange flowers are bird magnets, often providing so much nectar for their hungry visitors that this sugary treat drips from the flowers. The flowers themselves tend to have tubular shapes to accommodate bird beaks and tongues, and they are sturdy enough to withstand buffeting by these relatively heavy pollinators. Birds don't have much of a sense of smell, however, so bird-pollinated flowers generally lack fragrance.

Moths and butterflies also seek out red and orange flowers, but in contrast to birds they are strongly attracted by scents. The night-blooming flowers, like jasmine, don't waste energy on colors but are a luminous white with heady perfumes to attract nocturnal moths.

Bees and some other insects don't really notice red at all. Their best vision is down in the blue end of the spectrum, so blue and purple flowers like foxglove, lupine, and delphinium are bee attractors. Interestingly, bees also see into the ultraviolet region, which isn't visible to us. But many bee-pollinated flowers have ultraviolet guides, which act like miniature landing strips on sturdy petals, all leading to the center of the flower. White flowers, in particular, often look very

Bees can see ultraviolet light, which reveals floral landing strips. The photo on top shows what we see. The photo below was taken using an experimental filter that allows us to see what bees see.

Outfoxed by Foxglove

When I'm giving talks to gardeners I sometimes show a picture of a round purple flower that's heavily speckled and ask my audience what's wrong with the plant. Most people think it's a disease problem, but it's actually just another freaky flower phenomenon that's rooted in the past. Primitive flowers were radially symmetrical (round), much like current day sunflowers, magnolias, and roses. As insects and other airborne critters discovered the all-you-can-eat buffet of protein-rich pollen and sugary nectar, some flowers took advantage of this free shipping option for pollen by chang-

ing their shapes. Bilaterally symmetrical flowers (those that have right and left sides) are able to force pollinators into places where they are most likely to pick up pollen accidentally and transfer it elsewhere.

Foxglove has bilaterally symmetrical flowers and bees crawl into the tubular blossoms to forage for nectar. Every once in a while, however, a flat, disc-shaped flower will show up on a foxglove spike, usually at the terminal end. This phenomenon is called peloria and is simply a reversion to the primitive radial form. You can frequently find peloric flowers in orchids, and even Darwin found them in snapdragons.

This foxglove flower has reverted to a primitive form.

different under ultraviolet light and obviously are catering to bees as pollinators.

Of course, there are other garden pollinators, including bats, beetles, flies—and you! What's your favorite garden flower? The first brilliant yellow daffodil of spring? Fragrant David Austin roses? Masses of electric blue mophead hydrangeas? Exotic night-flowering jasmine? Simple, sophisticated calla lilies? Whether it's the fragrance, color, or shape that appeals to you, plants use that to their advantage. Even though many of our ornamental cultivars are sterile, any time we press our noses into one flower after another we're transferring pollen. Along with our penchant for dividing and sharing garden and house plants, humans are possibly some of the best reproductive aids in the botanical world.

SPREADING THE WEALTH

It's obvious that gardeners not only have green thumbs, but they're *under* green thumbs in terms of obeying their plants' reproductive bidding. But plants have one more difficulty to overcome: how to disperse all of those seeds that have formed after successful pollination. Like pollen, seed dispersal relies on natural forces as well as animals.

Some seeds are spread by water. The most massive seed of all, the coconut, is moved by ocean currents far from the shoreline where the mother palm stood. Plants growing in ponds and along stream banks can also take advantage of water currents to float their offspring away. Others set their seeds on the wind using various buoyant structures to keep the seeds aloft as long as possible. The parasol seeds of dandelions, the rotary blades of maple achenes, and the cottony billows of poplars allow the offspring to establish their own territory and lessen competition with their parents and siblings.

But wind and water are unpredictable methods of transportation. Seeds may end up in areas where they can't survive. Animals, in contrast, not only move seeds but often provide a nice dollop of fertilizer to go along with them. Therefore, many plants have developed tasty

The Hottest Thing
for Your Garden

Many gardeners recognize anthuriums, though only those in tropical climates can grow them outside. Other members of the arum family (Araceae) include calla lilies, jack-in-the-pulpit, skunk cabbage, philodendron, and the awesome titan arum. One thing these plants have in common is a spadix, which consists of a multitude of tiny flowers clustered on a sturdy stem. This suggestively shaped structure has given rise to some snicker-worthy scientific names, like *Amorphophallus titanum*, which means giant misshapen penis. Ahem.

Indeed, the spadix plays an unusual role in reproduction for many plants in the arum family. It produces heat—a lot of heat—through an unusual biochemical pathway. Some flowers have been measured to reach over 100°F. The heat helps volatilize odors that

Skunk cabbage flowers generate enough heat to melt snow.

Titan arums are visually stunning and equally malodorous.

attract pollinators, which unfortunately tend to be beetles and flies that like rotting meat. Skunk cabbage is an aptly named arum and one that explains why some arums aren't a popular gardening choice.

This heat production has other roles for arums in cold climates: it provides a haven for pollinators (like the heliotropic plants in the arctic) and allows arums to get a jump start on early spring growth by melting the surrounding snow and ice.

Fruit eaters know to avoid eating green fruit,
which is not yet ready for seed dispersal.

and nutritious wrappings for their seeds, as inducements to animals to serve once again as unwitting chauffeurs.

Plants use color and fragrance to advertise the edible rewards available for animals willing to take the kids for a ride. The hard green tissues of unripe fruits gradually plump, sweeten, and turn all colors of the rainbow, signaling to hungry animals that the kitchen's open and dinner is served. This ripening time is necessary to ensure that the seed, tucked safely away in its tough seed coat, has reached maturity and can leave the mother plant.

Species that produce relatively small seeds, from berries to apples to melons, are adapted to having their offspring ingested right along with the sweet pulpy fruit. The seed coat resists complete digestion, but it is etched by stomach acids before the seed is finally expelled, ready to germinate and use up those conveniently deposited nutrients. Other seeds might have their coats scratched (or scarified) by wind, water, sand, or cold temperatures. Still others, like some pine species, need trial by fire to pop open the cones where the seeds are trapped.

Plants protect their seeds from animals with good reason: they are highly nutritious, containing protein, carbohydrates, and often fats. Think nuts, legumes, and grains. So some plants, especially those with larger seeds, have turned to more sinister means of protection. Apricot pits contain chemicals that turn to cyanide, and castor beans contain ricin. Many of the unique poisons made by plants are used to defend their offspring from seed-eaters.

From the most primitive moss to the tallest conifer and to the most elegant orchid, plants have managed to use every body part and exploit animals as well as the forces of nature to ensure their spread into every environment, on every continent except Antarctica. So when you're in your garden this year, pulling those annoying spring weeds, enjoying those first summer blueberries, or dividing your irises in autumn, think about this amazing accomplishment by a life form that's literally rooted in place.

SUGGESTIONS FOR FURTHER READING

Chalker-Scott, L. *The Informed Gardener* (2008) and *The Informed Gardener Blooms Again* (2010). University of Washington Press, Seattle, Wash.

This set of myth busting books is a comprehensive guide for gardeners that will help them differentiate between good and not-so-good gardening advice.

Chalker-Scott, L. (ed.). 2009. *Sustainable Landscapes and Gardens: Good Science, Practical Application.* GFG Publishing, Yakima, Wash.

This multi-authored book, written for gardeners and landscape professionals, presents the most current and relevant science for caring for plants in managed landscapes.

Cranshaw, W. 2004. *Garden Insects of North America.* Princeton University Press, Princeton, N.J.

Dr. Cranshaw has created a beautifully illustrated guide to the insects gardeners are most likely to discover. This book will help you separate friend from foe and treat accordingly.

Dirr, M. A. 2009. *Manual of Woody Landscape Plants: Their Identification, Ornamental Characteristics, Culture, Propagation and Uses.* 6th ed. Stipes Publishing, Champaign, Ill.

Dr. Dirr's seminal publication on trees and shrubs for urban landscapes. Regardless of where you live, you will find valuable information on the best trees and shrubs to select for your home garden.

Dirr, M. A. 2011. *Dirr's Encyclopedia of Trees & Shrubs.* Timber Press, Portland, Ore.

This volume combines Dr. Dirr's encyclopedic treatments of hardy and temperate-climate trees and shrubs, with thousands of plants and color photographs.

Gillman, J. 2008. *The Truth About Garden Remedies: What Works, What Doesn't, and Why.* Timber Press, Portland, Ore.

My colleague Dr. Jeff Gilman presents the science behind a wide range of home remedies commonly recommended to home gardeners. His ability to translate garden science for the public is matched only by his informal, approachable writing style.

Gillman, J. 2008. *The Truth About Organic Gardening: Benefits, Drawbacks, and the Bottom Line*. Timber Press, Portland, Ore.
    Once again Dr. Jeff clearly and rationally discusses the science behind organic gardening. Regardless of where you fall on the organic continuum, you will appreciate his clear-headed, accessible approach.

Harris, R. W., J. R. Clark, and N. P. Matheny. 2003. *Arboriculture: Integrated Management of Landscape Trees, Shrubs and Vines*. 4th edition. Prentice Hall, Upper Saddle River, N.J.
    A practical, research-based guide to caring for woody landscape plants. Unfortunately, it's a little dated, but much of the information is still applicable.

Reich, L. 1997. *The Pruning Book*. Taunton Press, Newtown, Conn.
    Pruning expert Dr. Lee Reich explains proper pruning techniques, using clear illustrations that leave no room for doubt.

Robinson, T. 1991. *The Organic Constituents of Higher Plants*. 6th edition. Cordus Press, North Amherst, Mass.
    An incredible compendium of plant biochemicals, including chemical structures, for the real chemistry geeks in the gardening audience.

Taiz, L., and E. Zeiger. 2010. *Plant Physiology*. 5th edition. Sinauer Associates, Sunderland, Mass.
    This current plant physiology text has oodles of new information. It's heavy on the genetic and molecular side of things, so I would not recommend it for those suffering from science-phobia.

ACKNOWLEDGMENTS

Like my home landscape, this book has taken a few years to mature into something I'm proud to share with others. My fellow Garden Professor colleague and garden writer Dr. Jeff Gillman made the critical introduction between me and Timber Press. The staff at Timber has been patient, understanding, and professional in our work together, and I was thrilled to work with such a highly regarded press.

My heartfelt thanks go to the numerous gardeners, garden writers, garden photographers, and garden professors who assisted along the way, including Constance Casey and Cindy Riskin for their thoughtful comments on early drafts; and to the thousands of gardeners worldwide who ask the questions that prompted me to write this book.

Lastly, my gratitude to Jim, Charlotte, and Jack for their patience and humor as they alternately tolerated or soothed the frenzies surrounding my writing process. I am so fortunate to have you as my family.

PHOTO AND ILLUSTRATION CREDITS

PHOTOS

Alamy

© blickwinkel, page 33

© dk, page 31

Allen D. Owings, courtesy LSU AgCenter, page 148

Bert Cregg, pages 29 right, 193

Chris Farrow, page 151

Connie Ma, page 155

© Dr Schmitt, Weinheim Germany uvir.eu, page 211

Elizabeth McCoy, page 124

Encyclopaedia Britannica/UIG Via Getty Images, page 131

Flickr

Andrea_44, page 143

Ano Lobb, page 205 top left

Bemep, page 105

brewbooks, page 139 top right

brxo, page 139 below left

cuplantdiversity, page 35

David Stegall, page 195

Dendroica cerulea, page 139 below right

Evelyn Fitzgerald/Virens (Latin for greening), page 118

Franz_Franz, page 198 top

Jennifer Boyer, page 169 right

Mick E. Talbot, page 29 left

Nicholas A. Tonelli, page 153

Ryan Johnson, page 214

Tom Rulkens, page 205 bottom

Tony Austin, page 205 top right

Upupa4me, page 169 left

iStockphoto

bedo, page 166

Rolphus, page 159

Juliet Blankespoor, page 208

Michael Piwonka, page 108

Treecreeper Arborists Ltd, page 158

Wikimedia

Challiyil Eswaramangalath Pavithran Vipin, page 114

Chrumps, page 139 top left

Cwmhiraeth, page 64

Frank Vincentz, page 97

Galwaygirl, page 198 bottom

Hans Bernhard (Schnobby), page 117 below

Melbcity, page 215

Ragesoss, page 216

Rasbak, page 62

Sebastian Stabinger, page 24

Tangopaso, page 161

Walter Siegmund, page 104

Yummifruitbat, page 116 top

Zephyris, page 180

All other photos are by the author.

ILLUSTRATIONS

Dave Carlson, pages 16, 18, 36, 49, 86, 94, 122, 162, 165

Kate Francis, pages 92, 127, 144, 204

Laken Wright, pages 47, 88, 133

INDEX

JACK SCOTT

LINDA CHALKER-SCOTT has a Ph.D. in Horticulture from Oregon State University and is an ISA-certified arborist. She is Washington State University's extension urban horticulturist and an associate professor in the Department of Horticulture, with affiliate associate professor status at the University of Washington. She develops educational materials for home gardeners, certified arborists, restoration ecologists, pesticide applicators, and the nursery and landscape industry. At WSU, she is co-chair of the Garden Team, an interdisciplinary group that produces science-based extension publications for home gardeners. She is the author of three other books: the award-winning, horticultural myth–busting *The Informed Gardener* and *The Informed Gardener Blooms Again*, and *Sustainable Landscapes and Gardens: Good Science, Practical Application*. She has published extensively in the scientific literature as well as in popular magazines such as *American Nurseryman*, *Organic Gardening*, and *Fine Gardening*. She and three other academic colleagues host *The Garden Professors* blog and Facebook pages, through which they educate and entertain an international audience.